CLIMATE CHANGE AND ITS CAUSES, EFFECTS AND PREDICTION

THE ROLE OF EMISSIONS IN CLIMATE CHANGE POLICY

Climate Change and its Causes, Effects and Prediction

Additional books in this series can be found on Nova's website under the Series tab.

Additional E-books in this series can be found on Nova's website under the E-books tab.

Energy Policies, Politics and Prices

Additional books in this series can be found on Nova's website under the Series tab.

Additional E-books in this series can be found on Nova's website under the E-books tab.

CLIMATE CHANGE AND ITS CAUSES, EFFECTS AND PREDICTION

THE ROLE OF EMISSIONS IN CLIMATE CHANGE POLICY

HEATHER C. PERKINS

EDITOR

Nova Science Publishers, Inc.

New York

Additional color graphics may be available in the e-book version of this book.

LIBRARY OF CONGRESS CATALOGING-IN-PUBLICATION DATA
Available upon request

ISBN 978-1-61728-986-6

Published by Nova Science Publishers, Inc. ✢ New York

CONTENTS

PREFACE

Congress is considering proposals for market-based programs to limit greenhouse gas emissions. Many proposals involve creating a cap-and-trade program, in which an overall emissions cap is set and entities covered by the program must hold tradable permits-or-"allowances"-to cover their emissions. According to the Congressional Budget Office (CBO), the value of these allowances could total $300 billion annually by 2020. The government could either sell the allowances, give them away for free, or some combination of the two. This book examines the implications of different methods for selling allowances, given available information and the experiences of selected programs with regard to climate change policy options.

Chapter 1- Congress is considering proposals for market-based programs to limit greenhouse gas emissions. Many proposals involve creating a cap-and-trade program, in which an overall emissions cap is set and entities covered by the program must hold tradable permits—or “allowances”— to cover their emissions. According to the Congressional Budget Office (CBO), the value of these allowances could total $300 billion annually by 2020. The government could either sell the allowances, give them away for free, or some combination of the two.

Some existing cap-and-trade programs have experience selling allowances. For example, member states participating in the European Union’s (EU) Emissions Trading Scheme (ETS) have sold up to about 9 percent of their allowances, and the amount of auctioning is expected to increase significantly starting in 2013. In the United States, the 10 northeastern states participating in the Regional Greenhouse Gas Initiative (RGGI) have auctioned about 87 percent of their allowances.

This chapter is part of GAO's response to a request to review climate change policy options. This chapter describes the implications of different methods for selling allowances, given available information and the experiences of selected programs. GAO reviewed relevant literature and interviewed program officials from the EU and RGGI, economists, and other researchers. This chapter contains no recommendations.

Chapter 2- The European Union's (EU) Emissions Trading Scheme (ETS) is a cornerstone of the EU's efforts to meet its obligation under the Kyoto Protocol. It covers more than 10,000 energy intensive facilities across the 27 EU Member countries; covered entities emit about 45% of the EU's carbon dioxide emissions. A "Phase 1" trading period began January 1, 2005. A second, Phase 2, trading period began in 2008, covering the period of the Kyoto Protocol. A Phase 3 will begin in 2013 designed to reduce emissions by 21% from 2005 levels.

Several positive results from the Phase 1 "learning by doing" exercise assisted the ETS in making the Phase 2 process run more smoothly, including: (1) greatly improving emissions data, (2) encouraging development of the Kyoto Protocol's project-based mechanisms—Clean Development Mechanism (CDM) and Joint Implementation (JI), and (3) influencing corporate behavior to begin pricing in the value of allowances in decision-making, particularly in the electric utility sector.

However, several issues that arose during the first phase were not resolved as the ETS moved into Phase 2, including allocation schemes and new entrant reserves, and others. A more comprehensive and coordinated response by the EU has been made for Phase 3 with harmonized and coordinated rules being developed by the European Commission.

Chapter 3- Both the European Union's Emissions Trading Scheme (EU-ETS) and the U.S. Title IV sulfur dioxide (SO_2) program provide insights into regulatory issues that may face any future U.S. carbon market. From the initial operations of the EU-ETS, the 2006 price crash raised questions about the adequacy of market regulation. In particular, some suspect that information about allocations leaked before official publication, and that certain traders profited from this knowledge.

Title IV's longer trading history reveals two important trends: (1) an increasing trend toward diverse and non-traditional participants that is likely to continue under a carbon market; (2), an increasing use of financial instruments to manage allowance price risk that is likely to expand under a carbon market as a hedge against price uncertainty. Indeed, a carbon market may look more like other energy markets, such as natural gas and oil, than the somewhat sedate SO_2 allowance market.

Regulation of emissions trading would have to consider two kinds of fraud and manipulation: fraud by traders or intermediaries against other investors, and sustained price manipulation. Four agencies could have roles in the regulation of an emissions market, each with its own attributes that may contribute to effective regulation.

Chapter 4- This chapter examines a major component of the emission allowance distribution strategy proposed in two major legislative vehicles under consideration in the 111th Congress: H.R. 2454 (Waxman- Markey), which passed the House on June 26, 2009; and S. 1733 (Kerry-Boxer), which was ordered reported by the Senate Committee on Environment and Public Works on November 5, 2009. Both bills (among other provisions) would establish a cap-and-trade program that would distribute—in the program's early years—a substantial portion of allowance value to electricity local distribution companies (LDCs) for the benefit of electricity consumers. Considering the magnitude of allowance value directed to LDCs, the potential consequences and challenges associated with this approach raise fundamental policy issues.

Although an LDC allocation approach would be effective in mitigating cap-related electricity price increases at the household level, a closer examination of this strategy reveals potential concerns. Recent economic studies indicate that an electricity LDC allowance distribution strategy would alter the carbon price signal. The overall cost of the cap-and-trade program would increase if the price signal is channeled away from economic sectors with abundant low-cost abatement opportunities to economic sectors that have fewer such opportunities. In addition, if the price signal is dampened in one sector of the economy or for a particular subgroup of society, the signal may shift to other sectors or other groups, yielding unintended impacts. Several studies have estimated the efficiency loss. The range of estimates indicates that the magnitude of the efficiency loss due to LDC allocation would be uncertain. This uncertainty may pose a challenge for policymakers, when evaluating trade-offs between various allowance distribution strategies.

In: The Role Emissions in Climate Change Policy ISBN: 978-1-61728-986-6
Editors: Heather C. Perkins

Chapter 1

CLIMATE CHANGE: OBSERVATIONS ON OPTIONS FOR SELLING EMISSIONS ALLOWANCES IN A CAP-AND-TRADE PROGRAM*

United States Government Accountability Office

WHY GAO DID THIS STUDY

Congress is considering proposals for market-based programs to limit greenhouse gas emissions. Many proposals involve creating a cap-and-trade program, in which an overall emissions cap is set and entities covered by the program must hold tradable permits—or "allowances"— to cover their emissions. According to the Congressional Budget Office (CBO), the value of these allowances could total $300 billion annually by 2020. The government could either sell the allowances, give them away for free, or some combination of the two.

Some existing cap-and-trade programs have experience selling allowances. For example, member states participating in the European Union's (EU) Emissions Trading Scheme (ETS) have sold up to about 9 percent of

* This is an edited, reformatted and augmented edition of a United States Government Accountability Office publication, Report GAO-10-377, dated February 2010.

their allowances, and the amount of auctioning is expected to increase significantly starting in 2013. In the United States, the 10 northeastern states participating in the Regional Greenhouse Gas Initiative (RGGI) have auctioned about 87 percent of their allowances.

This chapter is part of GAO's response to a request to review climate change policy options. This chapter describes the implications of different methods for selling allowances, given available information and the experiences of selected programs. GAO reviewed relevant literature and interviewed program officials from the EU and RGGI, economists, and other researchers. This chapter contains no recommendations.

WHAT GAO FOUND

The method of selling emissions allowances can have significant implications for a cap-and-trade program's outcomes, and therefore, it is important that the method be chosen based on well-defined goals. Goals often cited by program officials and economists include: maintaining simplicity and transparency, maximizing participation, promoting economic efficiency, generating a price that reflects the marginal cost of reducing emissions, avoiding market manipulation, raising revenues, and minimizing administrative costs. According to program officials, it is important to identify goals prior to choosing a sales method, as tradeoffs may exist. Some goals may also be interrelated—for example, a simple and transparent design may boost participation and reduce the risk of market manipulation.

Once goals are identified, policymakers face a number of choices regarding the design of a sales mechanism. Existing programs have used different mechanisms to sell allowances, including direct sales through exchanges and auctions. EU officials described exchange-based sales as effective and easy to implement, although they and other economists questioned whether this approach would be suitable for selling a high volume of allowances. Program officials also reported that auctions, the more commonly used sales mechanism in the EU and RGGI, effectively distributed allowances to program participants. However, some economists noted that auctions are not "one size fits all," and should be designed to take into account market characteristics, such as the number of potential buyers.

Using auctions to sell allowances would entail a number of other design choices. For example, policymakers could decide to utilize existing auction infrastructure, such as that used in exchanges or government auctions, or develop a new platform. Choices must also be made regarding the auction format and other design elements.

- *Auction format:* The auction format determines, among other things, the price that winning bidders pay for allowances and the number of bidding rounds. To date, ETS and RGGI auctions have used a single round format in which each participant that bids above a certain price receives allowances at that price. Program officials expressed general satisfaction with this format, and economists noted that its relative simplicity may encourage participation. However, some economists also recommended that policymakers consider other formats as well, such as multiple-round auctions, given that experience with large-scale allowance auctions has been limited to date.
- *Other auction design elements:* Apart from the auction format, other elements may affect outcomes, including: participation requirements, the frequency and timing of auctions, measures that establish lower or upper limits on allowance prices, and rules governing auction monitoring and the reporting of results.

ABBREVIATIONS

CBO	Congressional Budget Office
CRS	Congressional Research Service
ETS	Emissions Trading Scheme
EU	European Union
GDP	Gross Domestic Product
NAP	National Allocation Plans
OECD	Organisation for Economic Co-operation and Development
PPP	Purchasing Power Parities
RGGI	Regional Greenhouse Gas Initiative

February 24, 2010

The Honorable Max Baucus
Chairman
Committee on Finance

United States Senate
Dear Mr. Chairman:

Concerns about the effects of climate change have led the Congress to consider legislation that would limit emissions of greenhouse gases nationwide. Elevated concentrations of greenhouse gases in the atmosphere as a result of human activities could increase global temperatures and affect ecosystems, agricultural production, infrastructure, and human health. Among greenhouse gases produced by human activity, carbon dioxide is emitted in by far the largest volume, mostly as a result of the combustion of fossil fuels for electricity, transportation, and industrial processes.[1] Many of the legislative proposals to limit greenhouse gas emissions would create a cap-and-trade program under which the government would place an overall cap on emissions and issue tradable permits. Entities covered by the program would have to surrender enough permits for all of their emissions at the end of specified time periods. Depending on the program, these "covered entities" may include power plants, oil refineries, and other manufacturing facilities. Each permit—known as an "allowance"—would represent a set quantity of greenhouse gas emissions, such as one metric ton.[2] Allowances could be purchased and sold, creating a market in which the price of emissions fluctuates with supply and demand.

As we testified in August 2009, the government has two main options for distributing allowances, the value of which could total hundreds of billions of dollars annually by 2020.[3] One option is to give allowances away for free, which would transfer their value to recipients and may compensate covered entities for costs incurred as a result of the program. Another option is to sell allowances, which would generate revenue that could be distributed in a number of ways—for example, tax cuts that improve economic efficiency or lump sum rebates to consumers. Selling allowances could also discourage efforts to gain free allowances through lobbying or other activities and help ensure that the price of emissions is the same for both new entrants and existing entities.[4] As a result, many experts we interviewed for a previous report suggested that a cap-andtrade program should maximize the share of allowances sold.[5] Existing cap-and-trade programs that regulate greenhouse gases have used two principal methods to sell allowances—sales on an exchange and auctions—and their experiences with these methods may offer valuable lessons as Congress considers establishing a cap-and-trade program.[6]

This chapter is one of four responding to your request for information on climate change policy options.[7] Our objective was to describe the implications of different options for selling emissions allowances in a capand-trade program, given available information and the experiences of selected programs. To address this objective, we reviewed and analyzed academic and professional literature produced by research organizations, academic institutions, environmental groups, and industry associations, including international research. We also analyzed information on two cap-and-trade programs that have sold allowances—the European Union's (EU) Emissions Trading Scheme (ETS) and the Regional Greenhouse Gas Initiative (RGGI), an initiative of 10 U.S. Northeast and Mid-Atlantic states. We collected information on these programs from EU member state officials, European Commission officials, RGGI program officials, academic literature, and research organizations. In addition to interviewing officials involved with RGGI auctions, we selected a nonprobability sample of five EU member states—Austria, Denmark, Germany, Ireland, and the United Kingdom—for in-depth study, based on a review of background literature and interviews with knowledgeable officials. This sample enabled us to assess allowance sales that exhibited variation in several key areas: the size of the allowance market, the share of allowances auctioned, the design of the sale, and the amount of revenue generated. While the sample allowed us to learn about many important aspects of, and variations in, the design of allowance sales, it was not intended to provide findings that would be generalizable to all allowance sales. We also met with ETS officials and other stakeholders to discuss allowance sales methods used in individual EU member states as well as preparations for future large-scale auctions.[8] Appendix I provides a more detailed description of our scope and methodology.

We conducted our work from December 2008 to February 2010 in accordance with all sections of GAO's Quality Assurance Framework that are relevant to our objectives. The framework requires that we plan and perform the engagement to obtain sufficient and appropriate evidence to meet our stated objectives and discuss any limitations in our work. We believe that the information and data obtained, and the analysis conducted, provide a reasonable basis for any findings and conclusions in this product.

BACKGROUND

Existing cap-and-trade programs that regulate greenhouse gases, such as the EU ETS, have experience in the sale of allowances. As we reported in

November 2008, the ETS began the first of its trading periods, or "phases," in 2005.[9] In Phase I, which ran from 2005 to 2007, member states were allowed to auction up to 5 percent of their allowances, with the remainder distributed to covered entities free of charge. The auctioning limit increased to 10 percent in Phase II, which is to run from 2008 to 2012. The EU's decentralized approach gives member states the authority to design and execute their own sales. While some member states chose to sell or auction a portion of their allowances in Phases I and II, the quantity sold has been a relatively small percentage of the overall quantity of allowances distributed (see appendix III for more information).[10] For Phase III, which begins in 2013, the EU decided to increase the amount of auctioning significantly, and as a result approximately half of all the allowances will be auctioned.[11] The EU is currently assessing various auction design options for Phase III and beyond—including holding centralized, EU-wide auctions—and plans to adopt an official auctioning regulation by June 2010.[12]

U.S. programs also offer experience in emissions allowance auctions. The federal government has auctioned allowances for the emission of sulfur dioxide under its Acid Rain Program since 1993, and the Commonwealth of Virginia auctioned allowances for nitrogen oxide emissions—a pollutant that contributes to the formation of smog—in 2004. More recently, in 2005, RGGI was created to regulate the carbon dioxide emissions of large fossil fuel-fired generators in participating states. RGGI has auctioned nearly 87 percent of emissions allowances issued under the program for 2009, and each of the six centralized auctions held since September 2008 has raised between $38 million and $117 million for programs to promote energy efficiency and renewable energy projects, among other uses. In addition to auctions for emissions allowances, the U.S. government has experience conducting other types of auctions, such as for government securities, surplus property, oil leases, timber harvests, and electromagnetic spectrum licenses. The Treasury Department's Bureau of Public Debt, for example, conducts more than 250 public auctions per year involving over $5 trillion in marketable securities.

In a cap-and-trade program for greenhouse gas emissions, covered entities and other interested parties will be able to buy allowances not only from the government, but also from participants in the secondary trading market. In the ETS, for example, allowances can be purchased through over-the-counter markets or on exchanges such as the European Climate Exchange in London or BlueNext in Paris. Secondary market trading can involve a range of intermediaries—including banks and brokers—and several types of allowance transactions can occur. "Spot" sales involve the immediate payment and delivery of allowances between two parties. Market participants may also trade "forward" or "futures" contracts, both of which allow for delivery of allowances at a later date.[13] Futures contracts may be attractive to covered entities that wish to secure an allowance price in advance and reduce uncertainty about future compliance costs. Because the value of futures contracts fluctuates based on the current market price of allowances, parties that do not have a compliance obligation under the program may also wish to purchase them as an investment.

THE METHOD OF SELLING ALLOWANCES IN A CAP-AND-TRADE PROGRAM CAN INFLUENCE PROGRAM OUTCOMES

The literature and programs that we reviewed present many options for the design of a mechanism to sell allowances under a cap-and-trade program. We drew two major observations from the literature and the capand-trade programs we reviewed. First, because elements of the mechanism's design may affect outcomes—such as the price of allowances obtained at auctions and the cost of the program—it is important that design choices align with an emissions trading program's goals.[14] Second, once a goal is chosen, policymakers have numerous choices regarding the sale of allowances, including whether to sell them on an exchange or use auctions. If policymakers choose auctions—as did the majority of countries participating in the ETS and the RGGI states—they must also make important design choices in the areas of format, participation requirements, frequency and timing, price controls, and rules for reporting and monitoring.

Establishing Clear Goals Is Critical

Program officials and economists suggested establishing clear goals to help guide the design and implementation of government allowance sales in a cap-and-trade program. Identifying priorities early is critical to developing an effective sales approach, as certain designs may better serve certain goals. Goals commonly cited by researchers and program officials include:

- *Simplicity and transparency*. Many economists and program officials recommended that allowance sales be simple and transparent for all participants. Sales should be guided by rules that are clear and understandable—both to participants and to the general public—to encourage participation, prevent discrimination, and ensure easy access to allowances. To that end, several economists and program officials recommended selecting an auction format that is easy to use and does not involve complicated bidding procedures. Reporting sales results in a public and timely manner can also help to create a transparent market.

- *Maximizing participation.* Ensuring sufficient levels of participation in allowance sales is critical, according to available information and program officials we interviewed. Participation fosters competition and limits opportunities for collusion. Economists and program officials also advised that sales should not discriminate against any one group of participants, whether by excluding them directly or indirectly, such as through high transaction costs. Participation can also be encouraged with a simple and transparent auction design.
- *Economic efficiency.* Economic literature suggests that efficiency is a key goal for allowance sales. In the case of allowance auctions, economic efficiency can be achieved if allowances are purchased by those who value them the most. A general measure of the efficiency of an auction, therefore, is its ability to generate bids that accurately reflect how much value a bidder places on the allowance. If efficiency is achieved, the resource—in this case, the right to emit greenhouse gases—is allocated to its highest-valued use. Efficiency may be affected by strategic bidding behavior or collusion if these activities artificially depress the price of allowances.
- *Facilitating price discovery.* Allowance sales may help facilitate price discovery—the process of determining a commodity's price based on supply and demand. Sales that successfully facilitate price discovery will generate an allowance price that accurately reflects the marginal cost of reducing emissions. That is, the price would reflect decisions by covered entities either to reduce their emissions or to purchase allowances to cover them, whichever is more cost-effective. Economists expect that this process of price discovery will prompt emissions reductions by those covered entities that can undertake them most cost-effectively. Without effective price discovery, the overall efficiency of the allowance market may be diminished, increasing the program's costs to the economy. According to several program officials involved in the ETS, the need for price discovery

depends on the volume of allowances sold under the program. If only a small percentage is sold, and the rest freely allocated, price discovery will be accomplished in the secondary market.[15]

- *Avoiding market manipulation.* According to economic literature, allowance sales should limit opportunities for participants to collude or engage in other forms of market manipulation. Collusion to depress allowance prices, if successful, could distort price signals, reduce revenues collected by the government from the sale of allowances, and cause participants and observers to question the fairness and transparency of the program. In addition, if allowance ownership were to become concentrated among a small group of participants in the secondary market, these participants could then withhold allowances from the market, driving up prices and impeding efficiency. However, several program officials and economists said it would be difficult to pursue such a strategy in the presence of a liquid market, since participants would have to acquire large shares of allowances both at individual auctions and in the secondary market. In addition, literature and economists we interviewed suggest that a U.S. cap-and-trade program for carbon dioxide allowances would likely attract a large number of participants. According to them, broad participation would prevent any single participant from gaining undue influence within the market, limiting opportunities for collusion and market manipulation.
- *Revenue generation.* Available literature suggests that policymakers could also aim to maximize the level of revenues collected from the sale of allowances. For example, policymakers could design a sale so that it is more likely to achieve high allowance prices. However, while maximizing revenue is a common goal in other government auctions of public assets, high allowance prices could increase the burden of a cap-and-trade program on covered entities or consumers of their products, which could erode political support for the program. Furthermore, some program officials noted that it would be much easier to accomplish this goal by increasing the stringency of the emissions cap.
- *Minimizing administrative and transaction costs.* Several economists and program officials recommended minimizing the administrative and transaction costs associated with allowance sales. Administrative costs are the time and resources governments spend designing and implementing allowance sales; available information suggests that most of these costs are incurred in the design phase of the program.

> Transaction costs, on the other hand, refer to costs incurred by participants in obtaining allowances—for example, costs associated with registering for an auction, developing a bidding strategy, and any bidding fees. According to available information, high transaction costs could discourage participation, and smaller entities in particular may face disproportionately high costs relative to the value of allowances they purchase. According to economic literature, the design of an allowance sale can have a significant impact on both administrative and transaction costs. For instance, weekly auctions may result in higher administrative and transaction costs than those held less frequently, although using existing infrastructure may help to minimize these costs. While several economists noted that any costs incurred by governments and participants would likely be minor compared to the value of the allowances, program officials said that keeping these costs low would help ensure ongoing support for the program.

According to available literature and economists we interviewed, identifying the goals of the sale in advance can help policymakers evaluate the likely effects of a given sales method. This is especially important given that trade-offs may result from decisions regarding the various aspects of auction design. For example, a method that increases revenue collected from allowance sales may not be the most economically efficient approach. Some goals may also be interrelated—for example, a simple and transparent design may boost participation and reduce the risk of market manipulation.

Many Design Options Are Available to Meet Established Goals

The literature and programs that we reviewed present many options for the design of a mechanism to sell allowances in a cap-and-trade program. Each option has implications that will help determine the extent to which a program meets its goals. At a high level, policymakers face a choice about using auctions or other types of sales to distribute allowances. Auctions, if used, would entail additional design choices.

Sales or Auctions

According to available literature, selling emissions allowances through an exchange enables buyers to purchase allowances from an electronic platform as they would stocks or other commodities. While sales on an exchange are less widely used for emissions allowance distribution than auctions, two ETS member states have had experience with exchange- based sales: Germany (in ETS's Phase II) and Denmark (in ETS's Phase I).[16] Germany's sales, directed by Germany's state-owned bank, took place through two major European exchanges—the European Climate Exchange and the European Energy Exchange.[17] Allowances were sold daily, according to rules specified by the government. The bank used the same process as other members of the exchange wishing to trade allowances, such that buyers could not distinguish the government from other participants in the secondary market. Denmark's sales method differed from Germany's in that it paid a fee to two private firms to sell allowances on European exchanges. Rather than instructing the two firms to sell allowances daily, as Germany did, Denmark structured the fees paid to the firms such that they received an incentive to sell when they judged price conditions to be most favorable.

German and Danish officials we interviewed expressed general satisfaction with sales on exchanges and cited their potential strengths in meeting certain goals. For example, Germany's goal was to match the average secondary market price as closely as possible without disrupting the market, according to officials. Information provided by Germany's state-owned bank shows that the allowance price yielded through the exchanges did in fact mirror the market price of allowances sold during the same time period. Several officials praised the efficiency of such sales and reported general satisfaction among participants. Other potential advantages of sales over auctions include:

- *Lower administrative costs*. Germany and Denmark used existing exchanges, which made it unnecessary to design and administer auctions.
- *Ease of use*. Some member state officials we interviewed said that sales on exchanges are simpler than auctions on separate platforms because many of the large companies affected by the ETS are already registered and active participants in the exchanges. According to German officials, this is one reason Germany continued using existing exchanges when it made the transition from sales to auctioning in 2010.

Despite these possible advantages, implementing such sales may prove challenging given the scale of a potential U.S. program. While Germany expects to sell or auction 40 million allowances annually in Phase II of the ETS—the highest volume among EU member states to date—legislation being considered by the U.S. Congress proposes initially auctioning over 1 billion allowances in 2012 alone.[18] On such a large scale, auctions may be more feasible or desirable than sales, according to several economists and program officials. For example, European officials said that if large volumes of allowances are sold in this manner, it may be difficult to ensure that all participants, including smaller entities, are equally able to buy allowances at the market price. Some program officials also expressed reservations that that if a high volume of allowances were sold through exchanges, the government would become the dominant seller in the secondary market and affect the price formation process. Moreover, one economist pointed out that no extensive studies had been undertaken on the performance of sales on exchanges, whereas auctions are well understood as a mechanism of distributing government assets.

Program officials reported that auctions, the more commonly used sales mechanism in the EU and RGGI, effectively distributed allowances to program participants, although several noted that allowance auctions have not yet been implemented on a large scale. According to economic literature and the economists and officials we interviewed, potential strengths of auctions include:

- *Price discovery*. As economists and officials have noted, the process of auctioning helps to establish the cost of emission reductions and maintain an allowance price that reflects that cost. Auctions enable a government to put allowances together in "batches" and sell them at predetermined times, and may help regulated entities make business decisions that incorporate the cost of compliance with emissions regulations. As one economist explained, auctions encourage covered entities to assess their marginal costs of emissions abatement, consider their allowance needs carefully, and bid accordingly. However, some economists and officials said that regardless of whether auctions or sales are used, price discovery will occur in the secondary market, once that market becomes established.
- *Simplicity*. Many of our interviewees cited auctions' simplicity as an advantage over sales because auctions are well-understood. RGGI officials reported that covered entities with a range of auction experience received training in the RGGI auction process and had no trouble familiarizing themselves with it.

- *Lower transaction costs.* Holding periodic auctions—weekly, monthly, or quarterly, for example—may decrease the per-transaction cost of buying and selling allowances, since buyers and sellers would not need to devote the time and resources necessary to participate in daily trading.
- *Transparency.* According to economists and officials, auctions take place under established rules and time frames, and thus convey clear information about when and how the government will sell allowances. According to European Commission officials, establishing a clear and predictable auction calendar can help inform market participants of the precise timing of volumes coming into the market, thereby avoiding unnecessary uncertainty and price volatility.[19] Sales on exchanges may be less transparent, since it is difficult to monitor who is selling allowances and at what time.

Program officials noted that auctions can also be administered through exchanges. For example, Germany decided to conduct auctions using an existing European trading exchange beginning in 2010. According to a program official, this approach allows them to draw on preexisting auction infrastructure and administrative processes, as allowance bids are subject to the same rules as other exchange transactions. The European Commission's for the draft auctioning regulation governing ETS auctions in Phase III and beyond also proposes using an exchange or other trading platform to vet participants and administer auctions.

Allowance auctions could also be administered in other ways. For example, policymakers could use existing government auction mechanisms—such as those used by the U.S. Treasury to auction securities—to auction allowances. Alternatively, policymakers could choose to create a new auctioning platform or hire contractors to administer auction processes. For example, RGGI opted to use a proprietary auctioning platform, run by a contractor with experience in administering auctions for energy commodities.

If auctions are used, several other design determinations must be made, including: format, timing and size, participation requirements, price controls, and monitoring and reporting requirements.

Auction Format

The choice of auction format can affect how well the auction aligns with predetermined goals, such as maximizing simplicity, avoiding market manipulation, and aiding in price discovery. We focused our analysis on two classes of auctions most commonly discussed in the context of emissions allowance auctions—"uniform-price" and "discriminatory-price" auctions.

Choosing an appropriate auction format involves answering two key questions:

- What price should winning bidders pay for allowances?
    - In uniform-price auctions, all winning bidders pay the same price for the items purchased. In the hypothetical example illustrated in table 1, 5 companies bid for 10 allowances in a uniform-price auction. Even though they made different bids, each company that bids above a certain price receives allowances at the same price. This "clearing price" is the highest price point at which all of the 10 allowances available would be sold.[20] In this example, companies A, B, and C each receive allowances at the clearing price of $3.50, although Company C receives only 2 of the 3 requested allowances because the number of bids at or above the clearing price exceeds the number of allowances available at this price. Companies D and E, which bid below the clearing price, receive no allowances.
- In "discriminatory-price" auctions, winning bidders pay different prices for allowances purchased at auction. In some discriminatory-price auctions, winning bidders pay the amount of their bid. For example, in the hypothetical auction presented in table 1, Company A would get its full share of requested allowances at $5 each, Company B would pay $4 each, and Company C would receive part of its request at $3.50 per allowance.

Table 1. A Hypothetical Uniform-price Auction of 10 Allowances

Name of company	Number of allowances requested	Price per allowance Requested (in dollars)
Company A	6	5.00
Company B	2	4.00
Company C	3	3.50 Clearing price
Company D	4	2.75
Company E	4	2.50

Source: GAO.

Note: There were 10 allowances available. Companies A and B received the full allowance request,Company C received a partial allowance request, and Companies D and E received no allowances.

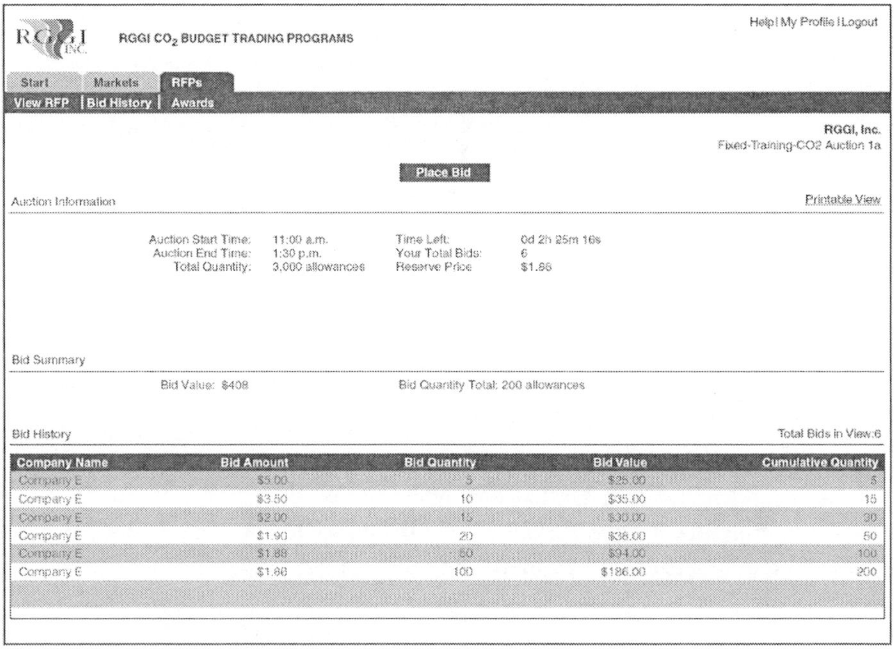

Company Name	Bid Amount	Bid Quantity	Bid Value	Cumulative Quantity
Company E	$5.00	5	$25.00	5
Company E	$3.50	10	$35.00	15
Company E	$2.00	15	$30.00	30
Company E	$1.90	20	$38.00	50
Company E	$1.88	50	$94.00	100
Company E	$1.86	100	$186.00	200

Source: Adapted from World Energy Solutions, Inc., RGGI's auction administrator. The table in the original replicates the RGGI auction bidding screen.

Figure 1. Participant's Bidding Screen, RGGI Uniform-Price, Single-Round Auction

- How many rounds of bidding should take place?
  - In a typical single-round auction, participants place bids once during a predetermined time period. Because participants do not see other bids before the outcome is announced, no opportunity is provided to change a bid based on information about others' bids. The single-round auction is thus sometimes referred to as a "sealed-bid" auction.[21] Figure 1 presents a simplified version of the RGGI software interface that participants use for RGGI's single-round auctions. As the figure shows, participants assemble a bid sheet, specifying the quantity of allowances requested at a stated bid price. Each participant may submit several bids at several different prices, if desired. According to a RGGI program

official, when RGGI's auction results are tabulated by the auction administrator's automated system, the results appear in a format similar to that shown in table 1. Participants whose bids appear above the line receive allowances, those bidding at the clearing price receive a partial allocation, and those below the line do not receive allowances.

- Auctions with multiple rounds of bidding occur in several formats, among them the "English" auction, in which the auction administrator raises the price of allowances round by round, and the "Dutch" auction, in which the auction administrator decreases the price round by round. Both the English and Dutch auctions are commonly referred to as "clock auctions," since the price is raised or lowered incrementally, like a clock's hands. Participants in multiple-round auctions have the opportunity to change the quantity of allowances for which they bid— or drop out of the bidding—as information is revealed round by round. Importantly, economists note that to discourage participants from potentially distorting allowance prices by increasing the quantity for which they bid in later rounds—after competitors have revealed their strategies—a clock auction can include a rule against increasing the bid quantity after the first bidding round.

In designing an auction, policymakers may consider selecting a format that is sensitive to the context of the allowance market. Previous U.S. federal government experience with auctions has involved different formats in different markets. For example, in 1994 the Federal Communications Commission chose to auction spectrum licenses in a simultaneous multiple-round format. By contrast, Environmental Protection Agency auctions of allowances to emit sulfur dioxide involve a single round where successful bidders pay as they bid, and auctions of government securities held by the U.S. Department of the Treasury involve uniform pricing.

In interviews and in economic literature, officials and economists have emphasized the importance of tailoring an auction for carbon dioxide allowances to the characteristics of the market, which may be different from other markets where auctions have been used. An auction of carbon dioxide allowances would sell many identical items—permits to emit a specified quantity of carbon dioxide in a particular time period.[22] Additionally, bidders in this market could include a large number of covered entities that emit carbon dioxide. These characteristics reveal both similarities and differences from some of the other markets listed above. For example, not all broadband spectrum licenses are alike, and their value to a buyer may further depend on the portfolio of licenses held. Furthermore, the number of potential buyers may be greater in the market for carbon dioxide emissions than sulfur dioxide emissions, in part because carbon dioxide is emitted in greater volume.

Existing cap-and-trade programs for carbon dioxide allowances—the EU ETS and RGGI—have employed the uniform-price, single-round format, in which winning bidders submit secret bids and pay the same price for allowances. Several program officials we spoke with expressed general satisfaction with this format, and the draft auctioning regulation governing ETS auctions in Phase III and beyond also proposes this approach. According to literature and economists we interviewed, advantages of this format include:

- *Simplicity.* For regulated entities that have participated in auctions, the simplicity and familiarity of the uniform-price, single-round format may prove valuable, according to several economists. This format has also proved easy to learn for those unfamiliar with auction processes, according to officials, as it involves relatively simple bidding procedures. One economist also reported that the uniform-price, single-round format is well-suited to automation compared to other auction formats, with much of the work handled by sophisticated but inexpensive computer programs. This economist pointed to RGGI, a small organization handling large pools of assets, as a case study in how simple the uniform-price, single-round auction can be to administer.
- *Avoidance of market manipulation.* Some economists said that other auction formats, such as clock auctions, may be more conducive to collusion than single-round auctions, because multiple bidding rounds give other bidders information and create opportunities for collusion.
- *Reduced risks for bidders.* Program officials and economists also said that the uniform-price, single-round format may alleviate concerns that could arise in discriminatory-price auctions. If a discriminatory-price auction requires participants to pay the value of their bids, for example, they run the risk of overbidding and paying more than other winning bidders for allowances. This may be of particular concern for small and inexperienced bidders, who may lack the information and resources to formulate a sophisticated bidding strategy. Uniform-price auctions reduce the possibility of making a costly bidding mistake, since all winning participants pay the same allowance price. For this

reason, some economists believe that uniform-price auctions will generate greater participation than discriminatory-price auctions.

Despite the strengths of the uniform-price, single-round format, some economists suggested that policymakers undertake further study before selecting an auction format. One study suggests that laboratory experiments with auction format options may provide insights that theoretical studies cannot, given the context-specific nature of the performance of various auction formats.[23] One economist also said that legislation need not specify a single auction format and could instead instruct government agencies responsible for the program to choose among various format options. Several RGGI states followed this path, by issuing regulations authorizing the auction administrator to use the uniform-price, single-round auction format or the ascending price, multiple-round format.[24] Policymakers could also leave room to revisit the auction format stipulated in cap-and-trade legislation, although introducing significant changes at later stages would require participants to relearn auctioning procedures.

Among auction format options other than the uniform-price, single-round format, the clock format may have comparative strengths in achieving certain goals, according to economists we interviewed. For example, economic literature suggests that clock auctions may lead to more reliable price discovery, since each participant may raise its bids in an attempt to win allowances, so that allowances go to those who are willing to pay the most.[25] However, economists who did experimental work on auction formats said that a clock auction fared no better in terms of price discovery than a uniform-price, single-round auction. Another argument for the clock format arises if multiple products are sold at an auction. For example, in addition to auctioning allowances for the current year, the government could decide to auction allowances of other future-year vintages—that is, allowances sold in advance of the compliance year(s) in which they may be remitted. A clock format would allow bidders to express preferences for different vintages, which may allow more readily for substitution of one vintage for another and prevent price irregularities.

The clock auction format may also present some disadvantages. An official involved with Ireland's auctions said they chose a uniform-price, single- round format after determining that a clock auction would be comparatively expensive and difficult to implement. The format may also complicate participation: one economist involved with Virginia's clock auctions of nitrogen oxide allowances received complaints about participants having to work at a computer terminal all day to compete in the auction. A single-round auction format, by contrast, would only require participants to submit a single bid sheet, similar to that shown in figure 1 above. The economist also pointed out that having thousands of participants monitoring a day of multiple-round auctioning in a large federal program would increase the cost of both participating in the program and administering it.

Other Auction Design Elements

Apart from format, policymakers would face a number of choices related to auction design, each of which has implications for program outcomes. Among other things, choices must be made regarding participation rules, the frequency and timing of auctions, the use of reserve prices or other price controls, and the monitoring and reporting of auction activities. We briefly describe each of these considerations below and provide additional detail in appendix II.

Participation. Maintaining high levels of auction participation can lead to greater competition which, in turn, can reduce the risk of collusion or other market manipulation. To maximize participation, economists and program officials recommended opening auctions up to as wide of a group of bidders as possible, including financial institutions and other entities that do not have compliance obligations under the program. According to them, limiting participation can increase the risk of market manipulation, making it difficult to ensure that all covered entities have access to allowances. In addition, program officials said that a well-designed vetting and registration system can reduce the risk that a participant will default on a bid.

Frequency and timing. According to economists, the frequency of auctions should be driven by the volume of allowances sold: higher volumes of allowances may require more frequent auctions. Available information suggests that holding frequent auctions, such as monthly or weekly, can help maintain market liquidity and provide flexibility to covered entities. On the other hand, some officials said frequent auctions may also complicate planning and increase administrative costs, depending on how the auctions are conducted. In terms of timing, many officials recommended auctioning future-year vintage allowances, which allow covered entities to secure allowances in advance and reduce the risks associated with fluctuating prices.

Price controls. Price controls could be implemented in a number of ways. A reserve price would set a price below which no allowances can be sold at an auction. Several program officials and economists suggested that setting a reserve price can be an effective way to guard against low auction clearing prices that may result from collusion or low participation. In addition, in some cases a reserve price may serve as a "price floor" throughout the secondary market. According to some economists and researchers, a price floor may help provide incentives for investment in low carbon technologies; however, some program officials cautioned that price floors could unduly interfere with the functioning of the allowance market. At the other end of the price spectrum, policymakers could also set upper limits on the price of allowances through price ceilings. While price ceilings could provide insurance against sustained high allowance prices, some program officials advised against the use of these measures, which they said could compromise emissions goals and impede international linkage of programs.

Monitoring and reporting. An effective system to report auction results and monitor activity can increase transparency and help oversight entities identify and correct instances of market abuse. Information on auction results can also provide information with which to evaluate and improve the program. As a result, several officials and economists recommended establishing a market monitor to track activity at auctions and in the secondary market. However, in reporting auction results, economists and officials cautioned against reporting certain information, such as bidders' identities, which they said could inadvertently facilitate collusion.

We conducted our work from December 2008 to February 2010 in accordance with all sections of GAO's Quality Assurance Framework that are relevant to our objectives. The framework requires that we plan and perform the engagement to obtain sufficient and appropriate evidence to meet our stated objectives and discuss any limitations in our work. We believe that the information and data obtained, and the analysis conducted, provide a reasonable basis for any findings and conclusions in this product.

As agreed with your office, unless you publicly announce the contents of this chapter earlier, we plan no further distribution until 30 days from the report date. At that time, we will send copies to the appropriate congressional committees and other interested parties. In addition, the report will be available at no charge on GAO's Web site at http://www.gao.gov.

APPENDIX I. SCOPE AND METHODOLOGY

Our review assesses the implications of different options for selling allowances in cap-and-trade programs. To address this objective, we first identified cap-and-trade programs that regulate carbon dioxide emissions and have sold allowances through auctions or other means. Programs that met these criteria were the European Union's (EU) Emissions Trading Scheme (ETS) and the Regional Greenhouse Gas Initiative (RGGI). We then selected a nonprobability sample of 5 member states involved in the ETS—Austria, Denmark, Germany, Ireland, and the United Kingdom—in order to assess different methods that EU governments have used to sell allowances.[26] This sample enabled us to assess allowance sales that exhibited variation in several key areas: the size of the allowance market, the share of allowances auctioned, the design of the sale, and the amount of revenue generated. While the sample allowed us to learn about many important aspects of, and variations in, the design of allowance sales, it was not intended to provide findings that would be generalizable to all allowance sales.

To identify various options for selling allowances, we identified and reviewed over 40 works of academic and professional research produced by economists, industry associations, research organizations, academic institutions, and environmental groups, including international research. We identified these works through an internet and database search using relevant key words such as "allowance sales" and "auction design." We also analyzed literature from government agencies, including the Congressional Budget Office (CBO) and the Congressional Research Service (CRS). Reviewing this research helped us to assess the different methods for designing an allowance sales mechanism and the potential implications of these methods. We did not independently assess the validity of data, assumptions, or methodologies underlying the economic studies we reviewed. We met with U.S. and international stakeholders including officials from RGGI and the European Commission as well as program officials in Austria, Denmark, Germany, Ireland, and the United Kingdom. We also conducted semistructured interviews with leading economists and researchers selected on the basis of their expertise in climate policy or auction design.

Appendix II. Other Elements of Auction Design

There are several key elements of auction design, in addition to format, that can affect whether an auction meets its established goals. This appendix provides more detail on the observations and recommendations of program officials, economists, and the economic literature regarding the following auction design elements: participation, frequency and timing, price controls, and reporting and monitoring.

Participation Requirements

Policymakers may shape participation in a cap-and-trade auction through various aspects of auction design. For example, auctions could be designed to restrict participation or to provide special assistance to smaller entities. Available information suggests that maintaining high rates of participation in allowance auctions can help promote liquidity and reduce the risk of collusion. We discuss three auction characteristics that help determine participation: participation limits, bid limits, and procedures for vetting and registration.

Participation limits. Available literature suggests that limiting or eliminating the ability of certain entities to participate in allowance auctions could reduce the amount of revenue generated through the auctions and hinder the efficient allocation of allowances. For example, excluding those entities that do not have a compliance obligation under the program would decrease overall auction participation, which, in turn, may increase the likelihood of collusive activities to depress allowance prices. In addition, if the auction clearing price falls below the price in secondary markets—whether due to collusion or other factors—auction participants may be able to buy allowances at auction and sell them on the secondary market for profit, thus capturing revenues that would have otherwise gone to the government. According to one study, the end result would be an implicit subsidy to the entities allowed to participate in the auction and a corresponding reduction in government revenue. Moreover, some program officials maintained that restricting certain groups from participating may present practical challenges given the interrelated nature of the marketplace—for example, while a buyer may not be a covered entity under the cap-and-trade program, it could be a parent company, supplier, or partner to a covered entity. According to researchers involved in the design of RGGI's auctions, attempting to assess and monitor these relationships could prove costly for an auction administrator.

Given these concerns, many economists and program officials favored maximizing the number of potential auction participants. Auctions held within the ETS and RGGI, for example, have allowed noncovered entities—including banks, brokers, and other private firms—to purchase allowances at auctions. In fact, noncovered entities have purchased between 16 and 30 percent of allowances sold in RGGI auctions thus far, according to auction results data published on RGGI's Web site. Economists and officials involved with these programs said that financial entities can play an important role in the market. For example, banks and brokers can foster liquidity and help provide regular price signals to covered entities. Smaller entities, in particular, may prefer purchasing allowances at financial institutions, as this relieves them of the need to learn the particulars of the auction process and develop an appropriate bidding strategy.

Bid limits. To prevent entities from hoarding allowances—which could allow these entities to gain a competitive advantage or raise the price of allowances, among other things—policymakers could set limits on the amount of allowances entities can purchase at auction. For example, in RGGI, associated entities can purchase no more than 25 percent of the allowances available in a given auction. However, such limits may be difficult to enforce, according to one program official, since one entity may be able to bid on behalf of another. Moreover, several economists and program officials we spoke with suggested that hoarding behavior would be highly unlikely in a future U.S. program, since an entity aiming to corner the market may have to buy the majority of allowances across several consecutive auctions, an unlikely possibility given the anticipated price and volume of allowances and the number of entities seeking them.

Vetting and registration. For an auction to be successful, participants must meet the financial commitments associated with their bids. Auctions may therefore include an application and screening process in which potential bidders demonstrate their eligibility to participate by providing information such as their credit and bankruptcy history.[27] The process could also include a declaration of "beneficial ownership," which would require bidders to declare whether they would bid on their own account or on behalf of another entity. According to available information, identifying the beneficiaries of allowance transactions may help the entity responsible for monitoring the market spot evidence of potential market manipulation. Economists and program officials also recommended requiring participants to post some type of financial assurance, such as a bond, deposit, or letter of credit demonstrating their ability to pay. Financial assurance requirements can serve as collateral in the event that participants are unwilling or unable to pay, and thus should be set at a level that ensures payment without discouraging participation. In Ireland's first auction, for instance, the nonrefundable deposit was set at about 3,000 Euro, or about $3,778. According to program officials, this amount was later determined to be insufficient to compel payment—for example, if the prices in the secondary market fell below the clearing price after the auction, it may have been less expensive for the bidder to simply forfeit their deposit and nullify the sale. As a result, Irish officials raised the deposit amount to about 15,000 Euro, or about $18,890, in the second auction.

While economists emphasize the importance of providing financial assurance, it is also important not to impose undue costs or paperwork requirements on participants. As a result, some economists recommended relying on established measures—such as credit scores—and simplifying the process as much as possible. An official involved with administering RGGI, for example, reported that RGGI made several improvements to its participant qualification procedures—including allowing electronic submissions and eliminating notarization requirements—after receiving feedback from participants. Program officials involved with the ETS also said that outsourcing vetting activities to an entity with experience in these activities, like an exchange, may help reduce the cost of these activities.

As existing programs have demonstrated, auction activities—including attracting and vetting potential auction participants and facilitating the bidding process—can be undertaken by either the government or a designated private entity. In RGGI, for example, a private consulting firm administers the auctions and conducts these activities. In contrast, Germany uses an existing emissions trading exchange to administer the auctions and conduct the required due diligence on potential auction participants. Another option to cut down on the government's administrative burden is to implement a "primary participant" model, an approach used in the United Kingdom to auction both government bonds and emissions allowances. In United Kingdom allowance auctions, all bidders must go through one of 11 registered primary participants—all large financial firms—to place their bids. The primary participants can also bid on behalf of themselves.[28] A similar approach is used by the U.S. government to auction Treasury bills and bonds—most are purchased by banks and securities brokers (primary dealers) that trade directly with the Federal Reserve, although individuals and other entities can also participate in the auction. According to a program official involved with United Kingdom auctions, a major advantage of this model is that primary participants are responsible for developing and implementing vetting procedures and assume the financial risks associated with default. In addition, the banks that serve as primary participants already have experience performing background checks.

However, a primary participant model may also have some drawbacks, according to program officials. For example, United Kingdom primary participants were initially not compensated for their services, which may have dampened participation in the first auction and resulted in a lower clearing price. Primary participants are now compensated for each indirect bid they facilitate by paying a discounted clearing price.[29] According to one official involved in United Kingdom auctions, this has improved auction participation and generated higher clearing prices. However, the official also acknowledged that some large bidders—such as electricity generators—oppose the primary participant model because it forces them to disclose their bidding strategy to another entity; they would rather participate in the auctions directly. As a result, the outline for the draft auctioning regulation for Phase III and beyond proposes allowing bidders to access auctions directly.

Provisions for smaller entities. Another option for encouraging participation is to enact special provisions for smaller entities, which may be discouraged from participating in auctions due to a lack of information or experience. For example, Austria and the United Kingdom have implemented "non-competitive" auctions, which allow some entities to bid for a limited amount of allowances prior to the actual (competitive) auction. By eliminating the need to compete with larger, better-informed entities, non-competitive auctions aim to provide assurance that smaller entities will be able to obtain needed allowances without running the risk of overbidding. However, thus far participation in non-competitive auctions has been low—of the 100,000 allowances set aside for Austria's first non-competitive auction, only about 5,000 were actually sold. Further, in the European Commission's 2009 consultation on the auction regulation, respondents showed little interest in incorporating noncompetitive auctions or other special provisions for smaller entities. While a Commission official we spoke with acknowledged that few small entities contributed to the consultation, this official and others reported that smaller entities have generally preferred to purchase their allowances from banks and brokers rather than at auction.

Frequency and Timing

If auctions are used to sell allowances, policymakers must consider issues related to timing, including how frequently to hold auctions and whether to auction future years' allowances in advance. According to available literature and economists we interviewed, the timing of auctions can have implications for market dynamics, prices, administrative costs, and participation. A variety of timing approaches are used in existing programs—for example, RGGI holds quarterly auctions, whereas Germany began holding two auctions twice weekly in 2010.

Program officials and economists we interviewed said that determining the appropriate auction frequency depends largely on the number of allowances auctioned during each year. Several noted that higher volumes of allowances may require more frequent auctions, so as to ensure a manageable and constant flow of allowances into the market. Because of uncertainty about the size of a future U.S. program, one program official was hesitant to recommend a specific frequency.

Nevertheless, available literature and our interviews point to several arguments in favor of holding auctions relatively frequently, such as weekly. First, economic literature indicates that frequent auctions can help maintain market liquidity and price stability. Because allowances would be sold in smaller batches, frequent auctions help encourage a constant flow of allowances into the market, reducing the impact of individual auctions on market prices. For this reason, the outline for the draft regulation outline governing Phase III of the ETS and beyond—when the level of auctioning is expected to increase significantly—proposes holding auctions at least weekly. Second, frequent auctions may help covered entities to meet their compliance obligations in a timely and flexible manner, rather than running the risk of submitting a losing bid and having to wait several months until the next auction. Third, frequent auctions may alleviate the need for a covered entity to set aside large amounts of capital to compete for bigger, less frequently available blocks of allowances, which may be especially difficult for smaller entities. Frequent auctions also may facilitate efficient and flexible transactions through financial or other intermediaries, which can benefit both small and large covered entities. Finally, smaller, more frequent auctions may help mitigate the risk that participants could purchase a substantial fraction of allowances in an attempt to manipulate allowance prices.

However, holding frequent auctions may present trade-offs, according to available literature. Administrative and transaction costs could rise if auctions are held more frequently, and higher costs could reduce participation. In addition, officials involved in administering RGGI's quarterly auctions said it would be difficult to conduct necessary pre– and post-auction activities—including finalizing sales, returning auction collateral to bidders, and compiling reports—if auctions took place more frequently. Another potential disadvantage of frequently held auctions is the risk that participation will be low at some auctions.

Policymakers may choose to sell future-year vintage allowances in advance of the compliance year(s) in which they may be remitted. For example, in addition to offering allowances for the current 3-year compliance period, RGGI auctions also offer participants the ability to purchase allowances for the second compliance period, which is to start in 2012. In the ETS, allowances for Phase III of the program, which begins in 2013, may be auctioned as early as 2011 or 2012, according to the draft outline for the auctioning regulation.

Available literature and economists and officials we interviewed identified several potential benefits associated with auctioning allowances prior to the compliance period in which they may be remitted. Most importantly, it enables covered entities to hedge against the uncertainty of future allowance prices by purchasing them in advance. Auctions of future-year allowances may be particularly beneficial to electricity generators, which often establish contracts for fuel and electricity one to three years ahead of delivery. Entities could also potentially hedge price risk by establishing futures contracts with financial intermediaries; however, these intermediaries may charge risk premiums that may be passed on to customers.

Available information suggests that holding auctions even before the cap-and-trade program's first compliance period may help to jump-start the process of price discovery and improve liquidity. For example, RGGI held its first auction in September 2008, approximately 3 months before the first compliance period began. A RGGI official said that this early auction provided price information that proved beneficial to financial institutions and covered entities alike.

However, holding advance auctions may also present risks. According to one program official, if actual emissions under the program are lower than expected, auctioning a greater number of allowances early on may depress prices in the short term.

Price Controls

In designing a cap-and-trade program, policymakers may decide to set limits around the price of emissions allowances sold at auctions or in the secondary market. For example, setting a reserve price would establish a minimum price below which no allowances could be sold at auction. The use of reserve prices is relatively common in greenhouse gas auctions—for example, RGGI and the EU member states we reviewed all used reserve prices.[30] Policymakers could also decide to limit the extent to which allowance prices could rise and fall in the secondary market through price "floors" or "ceilings".

Reserve prices and price floors. According to economic literature and program officials, a reserve price may have several benefits. For example, a reserve price could reduce incentives for collusion by limiting the profitability of collusive activities. A reserve price could also be used to safeguard against unusually low clearing prices at an auction due to low participation or other unforeseen events. According to one economist, an auction that produces allowance prices that are substantially lower than those in the secondary market may raise concerns about efficiency and equity. Protecting against such a scenario is the primary reason that the United Kingdom chose to institute reserve prices for its auctions, according to a program official. However, this official did not expect the reserve price ever to be triggered, since the likelihood of insufficient participation or collusion was extremely low.

Despite the fact that allowances are traded in secondary markets—where the government does not control prices—a reserve price may have effects that extend beyond the auction itself. In some cases, for example, the reserve price may effectively set a "price floor" for allowances throughout the secondary market. According to economic literature, the extent to which a reserve price serves as a marketwide price floor depends on several factors, including the share of allowances to be auctioned and the ability to purchase offset permits or other imported allowances. For example, a CBO report said that a reserve price could create a price floor if the government chose to sell a significant fraction of emission allowances, as opposed to distributing them for free.[31] A key benefit of a price floor, according to some economists and researchers, is that it provides more consistent financial incentives for investment in low-carbon and energy-efficient technologies that could potentially reduce compliance costs in the long run. By establishing a minimum price on emissions, a price floor could also result immediately in more intensive use of low- carbon energy sources or encourage consumers to choose goods and services that are less carbon-intensive.

While few economists and program officials disagreed with the use of auction reserve prices as a general protective measure, some expressed concern about using a reserve price to implement a marketwide price floor. For example, program officials involved in administering the ETS cautioned that price floors are unnecessary and can unduly interfere with the functioning of the allowance market. Accordingly, a European Commission official said that while an auction reserve price may be incorporated into the auction design, the primary law underpinning the ETS bars the use of price floors. Finally, some program officials warned that price floors could limit participation from certain entities—such as large banks—by reducing their opportunities for profit.

If a reserve price is used, policymakers would also need to consider what to do if it is triggered. Because no bids below the price would be accepted, some allowances at the auction would go unsold. One option for addressing unsold allowances is to retire them by removing them from the program entirely, an approach some researchers and program officials support as a way to help preserve the program's environmental integrity. Specifically, triggering the reserve price may indicate that the emissions cap is too generous; retiring allowances that remain unsold at the reserve price would effectively make the cap more stringent. Another option would be to "roll forward" any unsold allowances to the next auction, an approach used by the United Kingdom in allowance auctions. Economists describe this approach as administratively simple; however, when any unsold allowances are rolled over, as opposed to retired, future allowance prices could be lower, reducing incentives for emission reductions. A third option would entail placing the unsold allowances into a "contingency bank" and releasing them for sale at the next auction that closed at a price above a pre-identified trigger price. This approach removes unnecessary allowances from the program while demand is low but keeps allowance prices (and compliance costs) from rising as sharply as they otherwise would in subsequent periods when demand is high. However, one economist cautioned that managing the bank could introduce political risks. Importantly, each of these three options applies to a scenario in which a reserve price is applied at auction. If policymakers decided to implement a firm price floor that applied throughout the secondary market, the government would have to guarantee a minimum price to sellers in that market. In this case, triggering the floor price would indicate that the quantity of allowances offered for sale at the floor price exceeded the quantity demanded by market participants. To guarantee the minimum price, the government could buy back the excess quantity. However, this could create budgetary and other complications, as the government would not be able to anticipate market outcomes.

Price ceilings and strategic reserves. To protect against unexpectedly high compliance costs, some cap-and-trade proposals also involve setting an upper limit on the price of allowances, either through a price ceiling— often known as a "safety valve"— or by establishing a "strategic reserve" of allowances. A safety valve would give covered entities the opportunity to purchase an unlimited amount of additional allowances from the government at a predetermined price. In the event the allowance price rose higher than the safety valve price, covered entities could buy allowances from the government at the lower price rather than purchasing them on the market. If the safety valve was triggered and additional allowances released, however, emissions could rise beyond the level set by the initial cap and compromise the program's emissions goals. As an alternative option, policymakers could set a strategic reserve of allowances to be released only if the price threshold is reached. The key distinction from a safety valve approach is that the allowances in the reserve would eventually be paid back in some way, thus maintaining the integrity of the cap over time. If the allowances in the reserve were used, for example, the corresponding increase in emissions could be offset, such as by tightening emissions caps in future years.

Some economists and program officials have cited several possible advantages associated with price ceilings. Setting a maximum price for allowances would provide insurance against unexpected price spikes in allowances—or a sustained period of high prices—either of which could cause the price of consumer goods and services to rise. If allowance prices are higher than expected, for example, a price ceiling could limit the costs to businesses and consumers while new technologies are developed that may achieve reductions at less cost. Because it provides some parameters around the cost of the cap-and-trade program, a price ceiling could also reduce the risk that firms that are both energy-intensive and trade-intensive will face competitive pressures from industries in countries without comparable limits on greenhouse gas emissions. Finally, establishing price ceilings in advance may increase the likelihood that the program will endure through severe economic fluctuations, thus providing certainty to investors.

Several of the potential disadvantages cited for price floors also apply to price ceilings—namely, that they may interfere with market functioning and discourage participation by financial entities. In addition to these disadvantages, a safety valve could have negative environmental implications if emissions rise considerably higher than the established caps. Available information suggests that establishing a safety valve in a U.S. program could impede linkages with other cap-and-trade programs, which would allow participants under a U.S. program to trade allowances with other programs, such as the ETS. In theory, linking can enhance the cost-effectiveness of the participating programs by enabling covered entities to take advantage of differences in the cost of abatement options. However, establishing a safety valve in one program would have implications for other linked programs—for example, linked countries may not be able to ensure that their emissions would be below a required level in a given year. Additionally, a price ceiling could discourage investment in research and development to create new energy-efficient technologies by limiting future profits from their sale. Moreover, some officials reported that price floors and ceilings are unnecessary if the initial cap is set correctly, using accurate, current emissions data, which they considered to be a better strategy for regulating prices than using price controls to artificially manipulate the market.

Reporting and Monitoring

According to available literature and program officials, establishing a system to report and monitor auction activities is an important aspect of auction design. Effective reporting can increase the program's transparency and help participants make informed bidding decisions. In addition, monitoring auction results can help government agencies or designated private entities identify instances of market abuse and evaluate whether auctions have met established goals.

Reporting of auction results. If auctions are used, policymakers would need to consider how and when to report data on auction results. Several program officials recommended making auction results available immediately after the auction. According to one official, providing timely and accurate data on auction results can provide covered entities with information on costs to use as part of their strategic planning efforts. In determining which data to report, program officials recommended that the government disclose aggregated data such as the quantity of allowances sold, the number of participants (and the fraction that won allowances), and the clearing price. RGGI makes auction results data available through an online program called CO2 Allowance Tracking System, which allows the public to download emissions data and relevant auction results.

While providing timely aggregated data can serve a useful purpose, economic literature suggests that revealing too much information about auction results could inadvertently facilitate collusion or limit auction participation. For example, publicizing the names of auction winners and their respective purchases may enable entities to determine whether collusive agreements established prior to the auction were honored. In addition, program officials said that reporting the identity of bidders allows market participants to discern the patterns and strategies of others. According to a European Commission official, entities may choose not to participate at auctions if they fear that commercially sensitive information will be revealed through their bidding. Some program officials noted that while restricting access to such data may run counter to traditional ideas about transparency, it may serve the public's best interest to establish clear limits around the amount and type of auction data released.

Monitoring and oversight. An auction monitoring system can help identify cases of market abuse, ensure that auctions comply with established rules, and provide information useful in evaluating and improving auction design. Several program officials recommended designating a market monitor to observe and assess activities at auctions and in the secondary market. The auction monitor could either be a private or public authority—for example, the RGGI program uses a private consulting firm to perform a number of market monitoring activities. According to a RGGI official, these activities include: ensuring auction rules are consistently applied in each auction, analyzing participant behavior and identifying any irregularities, and modeling the impact of potential design modifications to the program. The RGGI market monitor also analyzes secondary market activity, although the Commodity Futures Trading Commission (CFTC) is responsible for protecting market participants against fraud, manipulation, and abusive trading practices.

In Phase I and Phase II of the ETS, individual member states have used different methods to monitor auction results. For example, the United Kingdom government manages auctions but has appointed an independent observer to ensure auctions are conducted in accordance with the law. In Austria, an energy exchange handles vetting procedures and auction administration while the government handles oversight. For Phase III of the ETS and beyond, the EU Directive requires member states to report to the

European Commission on various aspects of auction outcomes, including issues related to access, price formation, and technical issues. In addition, the outline for the draft auctioning regulation from Phase III and beyond foresees the appointment of a single auction monitor. Respondents to the EU's consultation on Phase III auctioning generally favored this proposal, although some respondents noted efforts to curb market abuse at auctions would be ineffective if these efforts are not accompanied by similar efforts in the secondary market.

APPENDIX III. ALLOWANCE SALES CONDUCTED IN THE EU ETS AND RGGI

Table 2. Allowance Sales in the EU Emissions Trading Scheme as of 12/31/09

Member state	Method of sale	Year of sale	Amount sold		Clearing price	Amount of revenue generated
			Percentage of each member state's cap sold (or expected to be sold)	Number of allowances sold (in millions) as of 12/31/09		
Ireland	Auctions (3 total) Sale (1 total)	2006 and 2008 (auctions) 2009 (sale)	2.3 (Phase I) 0.5 (Phase II)	1.7	Auction 1: $33.12 Au-ction 2: 8.65 Auction 3: 0.01 Sale: $17.94	$19.9 million
Hungary	Auctions (2 total)	2006 – 2007	2.5 (Phase I)	2.4	Auction 1: 9.34 Auction 2: 1.09	12.5 million
Denmark	Sales on an exchange	2006 - 2007	5.0 (Phase I)	5.1	Weighted average price: 2006-2007: $7.62	38.3 million
Lithuania	Auction (1 total)	2007	1.5 (Phase I)	0.6	Auction 1: $.07	40,915
United Kingdom	Auctions (7 total)	2008 - 2009	7.0 (Phase II)	29.0	Auctions 1 to 7: Between $12.87 and $19.36	472.7 million
Austria	Auctions (2 total)	2009	1.3 (Phase II)	.5	Auction 1: $13.66 Auction 2: $16.68	6.1 million
Germany	Sales on an exchange	2008 - 2009	8.8 (Phase II)	80.0	Weighted average price: 2008 sales: $27.97 2009 sales: $15.49	1.7 billion

Source: GAO analysis of ETS data, obtained through interviews with program officials and documents from the European Commission, member state governments, and ETS auction administrators.

Notes: Data as of December 31, 2009. For auctions held in Phase I, this figure represents the percentage of actual allowances sold; for auctions held in Phase II, it represents the projected amount of allowances to be auctioned, based on National Allocation Plans (NAP) submitted by each member state. The Phase II NAPs for the United Kingdom and Austria also indicate that unused allowances from the New Entrant Reserve will be auctioned; we did not include these allowances in our calculations. Purchasing power parities (PPP) published by the Organization for Economic Cooperation and Development (Main Economic Indicators, February 2010) were used to convert sales prices and revenues from euros to dollars. The gross domestic product index was used to adjust the resulting figures for inflation.

Table 3. Auctions Held in the Regional Greenhouse Gas Initiative (RGGI)

Monetary amounts in dollars					
Date of auction	**Vintage sold**	**Number of allowances auctioned (rounded)**	**Number of entities submitting bids**	**Clearing price**	**Amount of revenue raised (rounded)**
9/25/08	2009	12.6 million	59	3.07	$38.6 million
12/17/08	2009	31.5 million	69	3.38	106.5 million
3/18/09	2009	31.5 million	50	3.51	117.2 million
	2012	2.2 million	20	3.05	
6/17/09	2009	30.9 million	54	3.23	104.2 million
	2012	2.2 million	13	2.06	
9/9/09	2009	28.4 million	46	2.19	66.3 million
	2012	2.2 million	12	1.87	
12/2/09	2009	28.6 million	62	2.05	61.6 million
	2012	1.6 million	8	1.86	

Source: GAO analysis of RGGI auction reports issued by Potomac Economics, RGGI's independent market monitor.

End Notes

[1] The six primary greenhouse gases include carbon dioxide, methane, and nitrous oxide, as well as three types of synthetic gases: hydrofluorocarbons, perfluorocarbons, and sulfur hexafluoride.

[2] Programs that cover emissions other than carbon dioxide may quantify these emissions in terms of carbon dioxide equivalent. Carbon dioxide equivalents provide a common standard for measuring the warming efficiency of different greenhouse gases and are calculated by multiplying the emissions of the non-carbon dioxide gas by its global warming potential, a factor that measures its heat-trapping ability relative to that of carbon dioxide.

[3] GAO, *Climate Change Policy: Preliminary Observations on Options for Distributing Emissions Allowances and Revenue under a Cap-and-Trade Program*, GAO-09-950T (Washington, D.C.: Aug. 4, 2009). See also: *Auctioning under Cap-and-Trade: Design, Participation, and Distribution of Revenues Before the Senate Comm. on Finance*, 111th Cong. (2009) (statement of Douglas Elmendorf, Director, Congressional Budget Office (CBO)). CBO notes that the actual value of allowances would depend on the design of the cap-and-trade program.

[4] For example, a firm may increase its emissions in pre-regulation years to drive up its emissions baselines and increase its allowance allocation.

[5] GAO, *International Climate Change Programs: Lessons Learned from the European Union's Emissions Trading Scheme and the Kyoto Protocol's Clean Development Mechanism*, GAO-09-151 (Washington, D.C.: Nov. 18, 2008).

[6] We do not assess nongovernmental sales as part of this chapter. Nongovernmental sales include the secondary market and sales or auctions by noncovered entities that receive allowances from the government entity administering the cap-and-trade program. For example, proposed legislation allocates some allowances to local distribution companies, state natural resource agencies, Indian tribes and other noncovered entities, and authorizes them to sell, auction, or retire the allowances.

[7] The three previous products are: GAO, *Climate Change Trade Measures: Considerations for U.S. Policy Makers,* GAO-09-724R (Washington, D.C.: July 8, 2009); GAO, *Climate Change Trade Measures: Estimating Industry Effects*, GAO-09-875T (Washington, D.C.: July 8, 2009); and GAO-09-950T.

[8] For information on future large-scale auctions, see: http://ec.europa.eu/environment/climat/emission

[9] For more information on the EU ETS, see GAO-09- 151.

[10] At the time of this chapter, Germany had committed to sell 40 million allowances per year—or 8.8 percent of all the allowances it will issue in Phase II—the most of any EU ETS member state since the inception of the program. According to program officials, Germany will also sell 1 to 1.5 million allowances to cover administrative costs.

[11] Electric power and certain industrial installations were originally covered under Phase I of the program. The scope of the ETS was expanded in 2009 to include other installations and certain flights of aircraft operators from or to EU airports. In 2012 and subsequent trading periods, 15 percent of the total allowances available for distribution to aircraft operators will be auctioned.

[12] Any auction, whether conducted by a member state or a centralized platform, must be open to covered entities throughout the EU because the ETS directive, or law, requires that all covered entities have fair, full, and equitable access to auctions in Phase III and beyond.

[13] In forward contracts, the specific terms are negotiated between buyers, whereas futures contracts refer to standardized agreements available only through exchanges.

[14] The price of allowances is affected by a number of factors, including the stringency of the cap, energy prices, and weather conditions.

[15] In the EU ETS, sellers that participate in the secondary market include banks, brokers, covered entities, and investors.

[16] In 2010, Germany began auctioning rather than selling allowances.

[17] The European Climate Exchange trades two types of carbon credits: EU ETS allowances and credits issued through the Clean Development Mechanism, a carbon offset program established under the Kyoto Protocol. The European Energy Exchange operates market platforms for both types of credits, as well for power, natural gas, and coal.

[18] H.R. 2454, 111th Cong. (2009).

[19] In some cases, price volatility may convey information about changes in market fundamentals. For example, the release of verified data in Phase I of the EU ETS indicated that the overall cap was likely to exceed total emissions. Prices fell in response, which established that the market was functional.

[20] Different ways may exist for determining the clearing price in a uniform-price auction. For example, the price may be the lowest bid price at which all available allowances received a bid, or the highest bid that is rejected.

[21] While the terms "sealed-bid" and "single-round" are often used interchangeably, sealed-bid auction procedures may also be used in multiple-round formats. For example, in a sealed-bid, multiple-round auction, participants could submit a set of bids in advance of the auction and the bids would be revealed over time as the auction progresses.

[22] If greenhouse gases other than carbon dioxide are covered under the program, allowances would likely represent one unit of carbon dioxide equivalent.

[23] Burtraw, Dallas, Jacob Goeree, Charles A. Holt, Erica Myers, Karen Palmer, and William Shobe, "Collusion in Auctions for Emissions Permits," Resources for the Future Discussion Paper 08-36.

[24] However, some state regulations required the uniform-price, single-round format for the first auction, but then authorized the auction administrator to choose among (1) the uniform-price, single-round format; (2) the ascending-price, multiple-round format; or (3) other formats for subsequent auctions.

[25] Cramton, Peter and Suzi Kerr, "Tradeable Carbon Permit Auctions: How and Why to Auction not Grandfather," *Energy Policy* (30) 2002, pp. 333-345.

[26] In the ETS, individual member states chose to hold national auctions in Phases I and II. Because all states participating in RGGI chose centralized auctions, we did not select a sample of RGGI states.

[27] According to some program officials, creditworthiness may be less of a concern if bidders are required to post collateral covering 100 percent of their bids.

[28] Under United Kingdom Treasury rules, primary participants must prevent the disclosure of confidential information they receive from indirect bidders to their employees responsible for preparing or submitting bids on the primary participant's behalf.

[29] The United Kingdom allows primary participants to pay a discounted clearing price— 5 Euro cents less per allowance—while charging its indirect bidders the actual clearing prices.

[30] Ireland's third and final auction did not have a reserve price.

[31] CBO, *Policy Options for Reducing CO_2 Emissions* (February 2008).

In: The Role Emissions in Climate Change Policy ISBN: 978-1-61728-986-6
Editors: Heather C. Perkins

Chapter 2

CLIMATE CHANGE AND THE EU EMISSIONS TRADING SCHEME (ETS): LOOKING TO 2020*

Larry Parker

SUMMARY

The European Union's (EU) Emissions Trading Scheme (ETS) is a cornerstone of the EU's efforts to meet its obligation under the Kyoto Protocol. It covers more than 10,000 energy intensive facilities across the 27 EU Member countries; covered entities emit about 45% of the EU's carbon dioxide emissions. A "Phase 1" trading period began January 1, 2005. A second, Phase 2, trading period began in 2008, covering the period of the Kyoto Protocol. A Phase 3 will begin in 2013 designed to reduce emissions by 21% from 2005 levels.

Several positive results from the Phase 1 "learning by doing" exercise assisted the ETS in making the Phase 2 process run more smoothly, including: (1) greatly improving emissions data, (2) encouraging development of the Kyoto Protocol's project-based mechanisms—Clean Development Mechanism (CDM) and Joint Implementation (JI), and (3) influencing corporate behavior

---

* This is an edited, reformatted and augmented edition of a United States Congressional Research Service publication, Report R1049, dated January 26., 2010.

to begin pricing in the value of allowances in decision-making, particularly in the electric utility sector.

However, several issues that arose during the first phase were not resolved as the ETS moved into Phase 2, including allocation schemes and new entrant reserves, and others. A more comprehensive and coordinated response by the EU has been made for Phase 3 with harmonized and coordinated rules being developed by the European Commission.

The United States is not a party to the Kyoto Protocol. However, five years of carbon emissions trading has given the EU valuable experience in designing and operating a greenhouse gas trading system. This experience may provide some insight into cap-and-trade design issues currently being debated in the United States.

- The U.S. requires only electric utilities to monitor CO_2. The EU-ETS experience suggests that expanding similar requirements to all facilities covered under a capand-trade scheme would be pivotal for developing allocation systems, reduction targets, and enforcement provisions.
- In the U.S. debate continues on comprehensive versus sector-specific reduction programs; the EU-ETS experience suggests that adding sectors to a trading scheme once established may be a slow, contentious process.
- As with most EU industries, most U.S. industry groups either oppose auctions outright or want them to be supplemental to a base free allocation. The EU-ETS experience suggests Congress may want to consider specifying any auction requirement if it wishes to incorporate market economics more fully into compliance decisions.
- EU-ETS analysis suggests the most important variables in determining Phase 1 allowance price changes were oil and natural gas price changes; this apparent linkage raises possible market manipulation issues, particularly with the inclusion of financial instruments such as options and futures contracts. The EU will examine the matter in preparation for Phase 3. Congress may consider whether the government needs enhanced regulatory and oversight authority over such instruments.

OVERVIEW[1]

Climate change is generally viewed as a global issue, but proposed responses typically require action at the national level. With the 1997 Kyoto Protocol now in force and setting emissions objectives for 2008-2012, countries that ratified the protocol are implementing strategies to begin reducing their emissions of greenhouse gases.[2] In particular, the European Union (EU) has decided to use an emissions trading scheme (called a "cap-and-trade" program), along with other market-oriented mechanisms permitted under the Protocol, to help it achieve compliance at least cost.[3] The decision to use emission trading to implement the Kyoto Protocol is at least partly based on the successful emissions trading program used by the United States to implement its sulfur dioxide (acid rain) control program contained in Title IV of the 1990 Clean Act Amendments.[4]

The EU's Emissions Trading System (ETS) covers more than 10,000 energy-intensive facilities across the 27 EU Member countries, including oil refineries, powerplants over 20 megawatts (MW) in capacity, coke ovens, and iron and steel plants, along with cement, glass, lime, brick, ceramics, and pulp and paper installations. In addition, aviation is currently being phased into the ETS. These covered entities emit about 40%-45% of the EU's total greenhouse gas emissions, and almost two-thirds of them are combustion installations. The trading program does not cover either carbon dioxide (CO_2) emissions from the transportation sector (except aviation), which account for about 25% of the EU's total greenhouse gas emissions, or emissions of non-CO_2 greenhouse gases, which account for about 20% of the EU's total greenhouse gas emissions. A Phase 1 trading period ran between January 1, 2005, and December 31, 2007.[5] A Phase 2 trading period began January 1, 2008, covering the period of the Kyoto Protocol, and a Phase 3 has been finalized to begin in 2013.[6]

Under the Kyoto Protocol, the then-existing 15 nations of the EU agreed to reduce their aggregate annual average emissions for 2008-2012 by 8% from the Protocol's baseline level (mostly 1990 levels) under a collective arrangement called a "bubble." In light of the Kyoto Protocol targets, the EU adopted a directive establishing the EU-ETS that entered into force October 13, 2003.[7]

One objective of the second phase of the ETS is to achieve 3.3 percentage points of the 8.0% reduction required by the EU- 15 under the Protocol.[8]

The importance of emissions trading was elevated by the accession of 12 additional central and eastern European countries to EU membership from

May 2004 through January 2007. For the new EU-27, the overall ETS emissions cap is set at 2.08 billion metric tons of carbon dioxide (CO2) annually for the Kyoto compliance period (2008-2012).

The second phase Kyoto compliance stage of the ETS is built on the experience the EU gained from its preliminary Phase 1. The European Commission (EC) believes that the Phase 1 "learning by doing" exercise prepared the community for the difficult task of achieving the reduction requirements of the Kyoto Protocol. Several positive results from the Phase 1 experience assisted the ETS in making the Phase 2 process run smoothly, at least so far. First, Phase 1 established much of the critical infrastructure necessary for a functional emission market, including emissions monitoring, registries, and inventories. Much of the publicized difficulty the ETS experienced early in the first phase can be traced to inadequate emissions data infrastructure.[9] Phase 1 significantly improved those critical elements in preparation for Phase 2 implementation.

Second, the ETS helped jump-start the project-based mechanisms—Clean Development Mechanism (CDM) and Joint Implementation (JI)—created under the Kyoto Protocol. [10] As stated by Ellerman and Buchner:

> The access to external credits provided by the Linking Directive has had an invigorating effect on the CDM and more generally on CO_2 reduction projects in developing countries, especially in China and India, the two major countries that will eventually have to become part of a global climate regime if there is to be one.[11]

Third, according to the EC, a key result of Phase 1 was its effect on corporate behavior. An EC survey of stakeholders indicated that many participants are incorporating the value of allowances in making decisions, particularly in the electric utility sector, where 70% of firms stated they were pricing the value of allowances into their daily operations, and 87% into future marginal pricing decisions. All industries stated that it was a factor in long-term decision-making.[12]

However, several issues that arose during the first phase remained contentious as the ETS implemented Phase 2, including allocation (including use of auctions and reliance on model projections), new entrant reserves, and others. In addition, the expansion of the EU and the implementation of the linking directives created new issues to which Phase 2 has had to respond.[13] Based on lessons learned in Phase 1 and Phase 2, the EU has taken a substantially different approach to these issues in Phase 3 that is discussed later.

RESULTS FROM PHASE 1 AND 2

It is unclear to what degree the first phase of the ETS achieved real emissions reductions. Emissions are dynamic over time; a product of a country's population, economic activity, and greenhouse gas intensity.[14] To capture these dynamics, each Member State of the EU developed an emissions baseline from models that project future trends in the country's emissions based on these and other factors, such as anticipated energy and greenhouse gas policies.[15] During the first phase, the emissions goal was to put the EU on the path to Kyoto compliance—not actually comply with the Protocol (which wasn't necessary until the 2008-20 12 time period). Thus, countries developed "business as usual" baselines based on projected growth in emissions. Such a projected baseline suffers from two sources of uncertainty: data uncertainties, and forecasting uncertainties. On data, Phase 1 suffered from uncertainties with respect to data collection and coverage, in monitoring methods for historic data, and data verification. On projecting future emissions, Phase 1 faced uncertainties with respect to economic or sector-based growth rates. Fueled in many cases by over-optimistic economic growth assumptions, these uncertainties increased the probability of inflated business as usual baselines.[16]

The combination of these factors and modest reduction requirements resulted in the emissions allocations for the 2005-2007 trading period being higher than actual 2005 emissions.[17] This result raised questions about how much reductions achieved during Phase 1 were real as opposed to being merely paper artifacts. On the positive side, verified emissions in 2005 were 3.4% below the estimated 2005 baseline used during the allocation process. In addition, the allowance prices for 2005 stayed persistently high, suggesting some abatement was occurring and raising questions of "windfall" profits. As stated by Ellerman and Buchner:

> First, and most importantly, the persistently high price for EUAs [EU emissions allowances] in a market characterized by sufficient liquidity and sophisticated players must be considered as creating a presumption of abatement. It would be startling if power companies did not incorporate EUA prices into dispatch decisions that would have shifted generation to less emitting plants. There is plenty of anecdotal evidence that this was the case,

> and the prominent charges of windfall profits assume that the opportunity cost of freely allocated allowances was being passed on (without noting the implications for abatement). Similarly, it would be surprising if there were no changes in production processes that could be made by the operators of industrial plants.[18]

However, EU emissions allowances (EUAs) during Phase 1 did not maintain value. Phase 1 EUAs were basically worthless during the final six months of 2007. This decline in EUA prices at least partially reflected the general non-transferability of Phase 1 EUAs to Phase 2. Only Poland and France included limited banking in their Phase 1 implementation plans (called National Allocation Plans (NAPs)). The EC further restricted use of Phase 1 EUAs in Phase 2 with a ruling in November 2006.[19] As a result, excess Phase 1 EUAs were worthless at the end of 2007.[20]

One consequence of the non-transferability of Phase 1 EUAs is that prices for Phase 2 EUAs remained relatively firm until the 2008-2009 recession reduced demand, as indicated by Figure 1. Scarcity is critical for the proper functioning of an allowance market. As further indicated by Figure 1, during 2009, the market firmed up at a much lower level as participants assessed the impact of the recession on the demand for EUAs. This is a different response than the market had during Phase 1, and may reflect Phase 2 improvements in the system. In particular, the more predictable 2009 response may reflect the ability of the EC to certify Phase 2 NAPs using more verifiable baseline data than were available for Phase 1.[21] A major reason the EC rejected *ex post* adjustments[22] was fear that such adjustments would have a disruptive effect on the marketplace.[23] Phase 1 did not firmly establish this foundation of markets;[24] based on the Phase 2 EUA future's market, further market development appears to be occurring, although, like most commodity markets, it remains somewhat volatile at times.

While the environmental performance of Phase 1 may be disputed, the need for additional reductions to achieve Kyoto is not. For 2008, the EU-1 5 is estimated to be 6.2% below its base- year emissions, compared with an 8% five-year average reduction commitment under the Kyoto Protocol. As indicated by Figure 2, this represents a continuation of reductions by EU-1 5 over the past five years. However, as indicated by the pink line, the European Environment Agency (EEA) projects that the EU-1 5 existing measures are insufficient to reduce EU-1 5 emissions to their Kyoto requirements (represented by the purple line), resulting in a projected 6.9% reduction from baseline levels. To achieve the Kyoto target the EU projects further actions reductions by EU-15 countries (represented by the green line in Figure 2), resulting in an overall reduction of 8.5% compared with baseline levels.

In addition to domestic emission reductions, the EU has also projected additional reduction credits received by activities permitted under the Kyoto Protocol: (1) purchase of project-based credits by ETS participants and EU governments (e.g., Joint Implementation (JI) and Clean Development Mechanism (CDM) projects); and, (2) the use of carbon sinks. As indicated in Figure 3, these activities provide a credit on the EU-15 baseline of 4.6 percentage points. Thus, if the EU-15 maintains its current path, it would exceed its Kyoto commitment by about 3.5 percentage points (6.9% minus 3.4%). If its planned measures result in the projected 8.5% reduction below baseline levels, the overachievement of its Kyoto commitment would be 5.1 percentage points (8.5% minus 3.4%).[25]

The EU-27 as a whole does not have an emissions target comparable to the EU-15 bubble. By 2010, EU-27 emissions are projected at 9.6% below Kyoto baseline levels assuming current policies. This reduction is projected at 11.3% if additional measures are included. Currently, 24 of the 25 countries with reduction requirements are projected to meet their commitments under the Kyoto Protocol.[26] Only Austria is not projected to meet its requirements even with additional planned measures and the use of Kyoto mechanisms.[27]

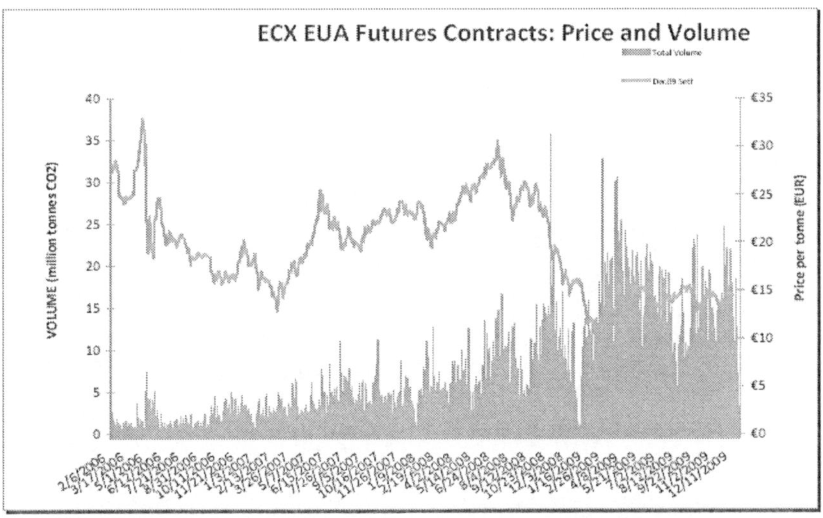

Source: ECX Exchange.

Note: Dec09 Sett: Future contracts with a settlement date of December, 2009.

Figure 1. ECX CFI Futures Contracts: Price and Volume

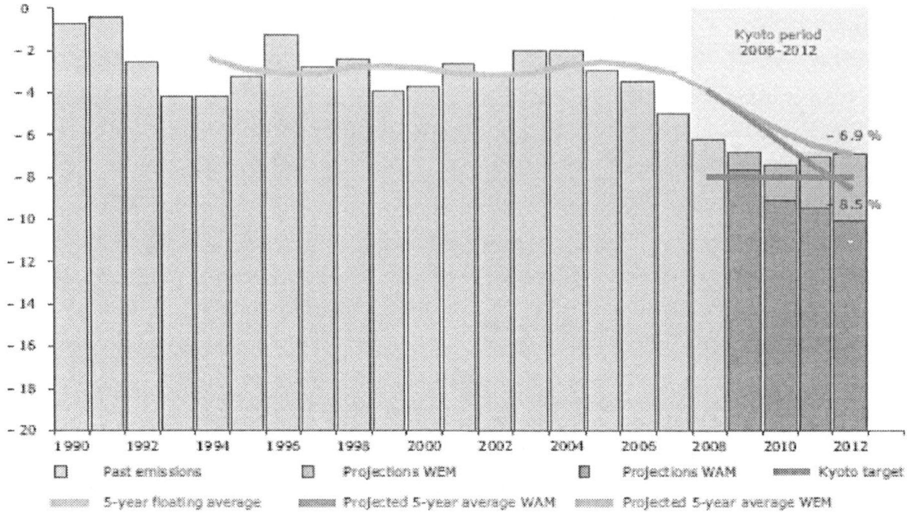

Source: European Environmental Agency, *Greenhouse Gas Emissions Trends and Projects in Europe 2000*, (2009) p. 10.

Note: WEM: with existing measures (measures implemented or adopted). WAM: with additional measures (planned measures).

Figure 2. EU- 15 Greenhouse Gas Emissions and Projections for the Kyoto Period: 1990-2012

PHASE 3

The European Union is committed to achieving a 20% reduction in greenhouse gas emissions by 2020 from 1990 levels (or more depending on the actions of other countries). A strategic component of the effort to achieve this target is a revised ETS that will achieve a 21% reduction from covered entities from 2005 levels. Table 1 indicates the proposed EU-wide ETS cap for the next Phase of EU greenhouse gas program (Phase 3) assuming no further international commitments (the final 2013 cap figure is required by June 30, 2010). As indicated, the EC envisions a linear reduction in the ETS cap to match the reduction target under the overall 20% reduction program. These numbers will change as individual countries decide to include more facilities under the ETS and as the EC expands ETS coverage to include other sectors and nonCO_2 greenhouse gases.

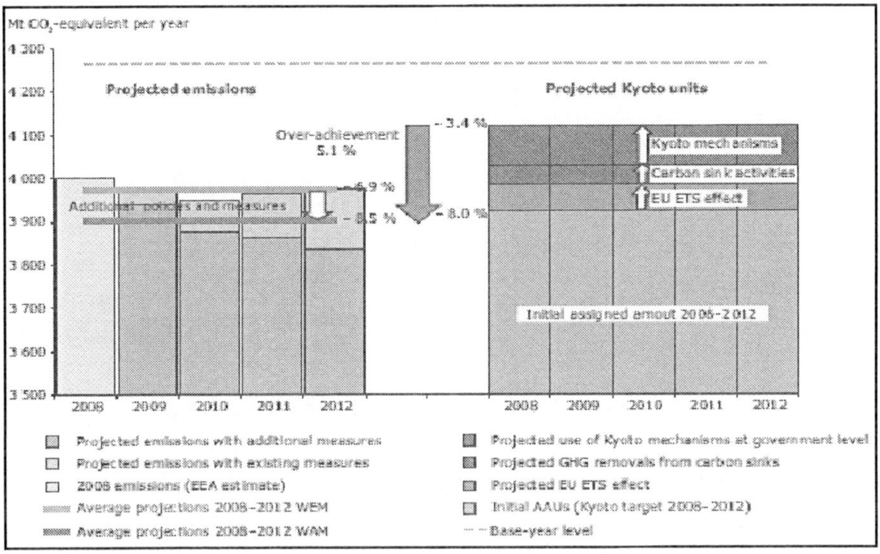

Source: European Environmental Agency, *Greenhouse Gas Emissions Trends and Projects in Europe 2000*, (2009) p. 11.

Notes: The left section shows the projected emissions considering only domestic measures (existing and additional) and is showing them as average 2008-2012 emissions (lines) and annual emissions (bars). The right section shows the projected amount of Kyoto credits accumulated by the end of the commitment period, including the initial EC assigned amount under the Protocol, the purchase of Kyoto project credits by EU ETS participants and EU governments, and carbon sink activities.

Figure 3. Summary of EU- 15 Emissions Projection Compared to Projected Kyoto Protocol Credits

For Phase 3, the EU is re-shaping the ETS to improve its efficiency and eliminate some of the problems identified during Phase 1 and 2.[28] For Phase 2, the improved emissions inventories resulting from Phase 1 allowed the EC to harmonize the types of installations covered by the ETS across the various Member States.[29] In addition, the EC imposed a uniform rule on the Member States preventing the use of *ex-post* adjustments. However, Phase 2 made little advancement in harmonizing individual countries' allocations schemes.[30] As with Phase 1, countries continue to differ widely on several key points.

The critical structural change the EU would make in Phase 3 is eliminating National Allocation Plans (NAPs), and replacing them with EU-wide rules with respect to allowance availability, allocations, and auctions. NAPs are central to the EU's effort to achieve its Kyoto obligations under

Phase 2. Each Member of the EU submitted a NAP that lays out its allocation scheme under the ETS, including individual allocations to each affected unit. These NAPs were assessed by the EC to determine compliance with 12 criteria delineated in an annex to the emissions trading directive.[31] Criteria included requirements that the emissions caps and other measures proposed by the Member State were sufficient to put it on the path toward its Kyoto target, protections against discrimination between companies and sectors, and delineation of intended use of CDM and JI credits for compliance, along with provisions for new entrants, clean technology, and early reduction credits. For the second trading period, the NAP had to guarantee Kyoto compliance.

This NAP structure will be replaced under Phase 3. There would be one EU-wide cap instead of the 27 national caps under Phase 1 and 2. Allowances would be allocated under EU-wide, fully harmonized rules, including those governing: (1) auctions, (2) transitional free allocations for greenhouse gas intensive, trade-exposed industries, (3) new entrants, and (4) coverage. The EC proposed a Directive to alter the EU-ETS structure for Phase 3 in January, 2008,[32] and the Directive was amended and adopted by the European Parliament (EP) and of the Council of the European Union in April 2009.[33]

Table 1. Proposed Annual ETS Cap Figures for Phase 3

Year	Billion metric tons of CO_2e
Annual limit for Kyoto compliance period (2008-2012)	2.083
2013	1.974
2014	1.937
2015	1.901
2016	1.865
2017	1.829
2018	1.792
2019	1.756
2020	1.720

Source: European Commission, *Questions and Answers on the Commission's Proposal to revise the EU Emissions Trading System* (Brussels, January 23, 2009), response to question 12.

Note: Figures are based on the current Phase 2 scope of the ETS. These need to be adjusted for three reasons: (1) extensions of ETS scope during phase 2 by Member states; (2) extensions of ETS scope by the EC for third trading period, and (3) the figures do not include inclusion of aviation, nor the emissions from Norway, Iceland, and Liechtenstein—non-EU countries that have linked their programs to the ETS.

Auctions

Under Phases 1 and 2, allowances generally were and are allocated free to participating entities under the ETS. During Phase 1, The EU-ETS Directive allowed countries to auction up to 5% of allowance allocations, rising to 10% under Phase 2.[34] Under Phase 1, only 4 of 25 countries used auctions at all, and only Denmark auctioned the full 5%. The political difficulty in instituting significant auctioning into ETS allowance allocations is the almost universal agreement by covered entities in favor of free allocation of allowances and opposition to auctions.[35] Free allocation of allowances represents a one-time transfer of wealth to the entities receiving them from the government issuing them.[36] The resulting transfer of wealth has been described by several analysts as "windfall profits."[37] As summarized by Ellerman and Buchner: "Allocation in the EU ETS provides one more example that, notwithstanding the advice of economists, the free allocation of allowances is not to be easily set aside."[38]

Despite concerns about windfall profits and economic distortions resulting from the free allocation of allowances, there was little change in basic allocation philosophy for Phase 2. No country proposed auctioning the maximum percentage of allowances allowed (10%). Most do not include auctions at all.[39] The unwillingness of governments to employ auctions as an allocating mechanism revolve around equity considerations, including: (1) inability of some covered entities to pass through cost because of regulation or exposure to international competition; (2) potential drag on a sector's economic performance from the up-front cost of auctioned allowances; and (3) the potential that government will not recycle revenues to alleviate compliance costs, international competitiveness impacts, or other equity concerns, resulting in the auction costs being the same as a tax.[40]

This opposition is mostly overcome for Phase 3 through an EU-wide set of harmonized rules for allowance allocations and auctions. Under Phase 3, the Directive states:

> Auctioning should ... be the basic principle for allocation, as it is the simplest, and generally considered to be the most economic efficient system. This should also eliminate windfall profits and put new entrants and economies growing faster than average on the same competitive footing as existing installations. (paragraph 15)

After nine eastern European Member States threatened to veto an initial proposal to auction 100% of all allowances, the EU compromised to provide for some free allocation of allowances during Phase 3 that will begin in 2013.[41] Most covered industries will be eligible for some free allocation of allowances to cover direct emissions under the Phase 3 agreement. The introduction of auction would be differentiated by sector. In general, for the power sector, full auctioning will begin in 2013. For electric powerplants, most will receive no free allocation of allowances during Phase 3. However, in a concession to certain eastern European Member States, an optional and temporary derogation from the no-free-allocation requirement for powerplants is provided to countries that meet specific energy and economic criteria. Under the optional allocation scheme, the Member State can allocate allowances equal to 70% of the powerplant's Phase 1 emissions free; this allocation declines to zero in 2020.

The auction schedule for most other covered entities is more gradual with 80% of a sector's allocation provided free in 2013, declining linearly to 30% by 2020, and zero by 2027. As stated in the final Directive:

> For other sectors covered by the Community scheme, a transitional system should be foreseen for which free allocation in 2013 would be 80% of the amount that corresponded to the percentage of the overall Community-wide emissions throughout the period 2005 to 2007 that those installations emitted as a proportion of the annual Community-wide total quantity of allowances. Thereafter, the free allocation should decrease each year by equal amounts resulting in 30% free allocation in 2020, with a view to reaching no free allocation in 2027. (paragraph 21)

Because of concern that stringent EU carbon policies may encourage production and related greenhouse gas emissions to shift to countries without carbon policies (i.e., carbon leakage), exceptions to this phase-out of free allowances will be made in sectors where carbon leakage may occur, as discussed later.

Distribution of allowances to be auctioned by the Member States will be determined by a three- part formula (Article 10(2)). Eighty-eight percent of the allowances to be auctioned by each Member State is distributed to States according to their historic emissions under Phase 1 of the EU-ETS. Ten percent of the total is distributed to States mostly based in their comparative GDP per capita within the EU (Annex IIa). Two percent of the total is distributed to nine former eastern-bloc countries based on the substantial greenhouse gas reductions they have already achieved (Annex IIb). Auctions will be conducted at the Member State level (cooperative auctions between

States are also allowed) and must be open to any potential buyer. The EC is directed to develop the appropriate rules for coordinated auctions by June 30, 2010.

Beyond the allocation of allowances, the EU Directive also provides guidelines for the allocation of revenues from allowance auctions. The Directive states that at least 50% of the proceeds should be used to fund a variety of climate change related activities, including emission reductions, adaptation activities, renewable energy, carbon capture and storage (CCS), the Global Energy Efficiency and Renewable Energy Fund, and assisting developing countries to avoid deforestation and increase afforestation and reforestation (Article 10(3)).

New Entrant Reserves

Unlike previous cap-and-trade programs, the EU-ETS includes provisions for allocating free allowances to new entrants to the system.[42] The reasoning behind this decision is based on equity: (1) it isn't fair to allocate allowances free to existing entities while requiring new entrants to purchase them, and (2) the EU doesn't want to put Member States at a disadvantage in competing for new investments.[43] These equity concerns trumped concerns about economic efficiency.

As is the case for existing entities, the free allocation of allowances to new entrants is a subsidy. Under Phase 1 and Phase 2, the size and distribution of this subsidy is left to the individual Member States. For Phase 1, the reserve varied widely from the average of 3% of total allowances: Poland set aside only 0.4% of its allocation for new entrants while Malta set aside 26%. For Phase 2, the spread continues with Poland reserving 3.2% of its allowances for new entrants in contrast to 45% reserved by Latvia.[44]

The decision to employ a new entrant reserve adds complexity to Member States' allocation plans and influences the investment decisions of covered entities. Rules had to be promulgated with respect to the reserve's size, manner in which the allowances are dispensed, and how to proceed if the demand either exceeds the supply, or vice versa. As indicated, countries did not harmonize new entrant reserve rules with respect to size during Phase 1 or 2. Likewise, there is no standardization on dispensing allowances and replenishing the reserve: first-come, first-serve with no replenishment is one approach used, but a variety of procedures have been developed both to dispense allowances and to replenish the reserve if supply is inadequate.

Member States also have different formulas for determining how many allowances a new entrant should receive. Member States claim to use a form of "benchmarking" to determine allowance allocations—an approach based on a standard of "best practices" or "best technology" that is applied to the new entrant's anticipated production or capacity. However, the definitions and application of the benchmarks used by the Member States are not uniform.

This will change under Phase 3. Under Phase 3, the Directive sets an EU-wide cap of 5% of the total allowance cap for a new entrant reserve, and requires the harmonization of allocation rules. The EC is to adopt a harmonized rule for applying a new entrant definition contained in the Directive by December 31, 2010; the Directive expressly excludes any new electricity production from being defined as a new entrant. The EC is also to determine EU-wide benchmarks for the allocation of all free allowances. The Directive states that the starting point for setting those benchmarks shall be the average performance of the 10% most efficient installations in a sector or subsector in the EU in the years 2007-2008 (Article 10a(2)).

In an attempt to stimulate development of CCS, the Directive also provides that up to 300 million allowances in the new entrants' reserve shall be available through 2015 for aiding construction and operation of up to 12 demonstration projects. No one project can receive more than 15% of the allowances allocated for this purpose (Article 10a(8)).

EC Phase 3 Decision on Eligible Industries[45]

Most studies of the competitiveness impacts of the ETS during Phase 1 have found no impact. The International Energy Agency (IEA) cites several reasons for this situation:

> Experience to date with the EU-ETS does not reveal leakage for the sectors concerned— analysis of steel, cement, aluminum and refineries sectors reveals that no significant changes in trade flows and production patterns were evident during the first phase (2005-2007) of the EU-ETS. This is mostly due to the free allocation of allowances, sometimes in generous quantities, and to the still functioning long-term electricity contracts, which softened the blow of rising electricity prices. Further, the general boom in prices for most traded products subject to carbon costs—whether direct or indirect—has blurred any effects of the latter. Finally, the relatively short time span of these policies does not allow observation of the full potential effects on industry via changes in investment location decisions.[46]

This conclusion is echoed by Carbon Trust, which states that currently, free allocation of emissions allowances offset almost all of the additional costs of the ETS; and that conclusion is echoed by The Climate Group for The German Marshall Fund, which states that companies surveyed found it difficult to quantify effects on their bottom line in the first phase, or found no effect at all.[47]

For energy-intensive, trade-exposed industries, Phase 3 has provisions to provide assistance to eligible installations to address the direct and indirect impact of emissions control costs. With respect to direct emissions costs, the EC published a list of installations exposed to a significant risk of carbon leakage on December 24, 2009, as required under the Directive.[48] The list is identical to the draft list released in September 2009.[49] The decision lists 164 industrial sectors and subsectors deemed to be exposed sectors under the appropriate European Parliament and Council directives. Eligible installations will receive allowances sufficient to cover 100% of their direct emissions, provided they are using the most efficient technology available. This 100% allocation contrasts with the 80% distribution of free allowances to non-carbon leakage exposed industries in 2013. Reflecting the fluid nature of the competitive situation and international negotiations, the EC is to review its decision June 30, 2010, and provide the European Parliament and Council with any appropriate proposals to respond to the situation.

Assistance for the impact of indirect emissions control costs on exposed industries from higher electricity prices would be determined by Member States. As stated by the Directive:

> Member States may deem it necessary to compensate temporarily certain installations which have been determined to be exposed to a significant risk of carbon leakage related to greenhouse gas emissions passed on in electricity prices for these costs. Such support should only be granted where it is necessary and proportionate and should ensure that the Community scheme incentives to save energy and to stimulate a shift in demand from grey to green electricity are maintained. (paragraph 27)

Flexibility Mechanisms and Price Volatility Control

The major flexibility mechanism developed under the EU-ETS has been the Clean Development Mechanism (CDM) and Joint Implementation (JI) credits permitted under the Kyoto Protocol; however, this development has proven a controversial process. A major part of the controversy has been the "supplementarity" requirement of the Kyoto Protocol to use its flexibility mechanisms. Supplementarity requires that developed countries, such as most EU countries, ensure that their use of JI/CDM credits is supplemental to their own domestic control efforts. In defining supplementarity for Phase 2, the EC used 10% of a country's allowance allocation as a rule of thumb in approving NAP s—with a greater limit possible based on a country's domestic efforts to reduce emissions. This process resulted in some significant reductions in some countries' proposed limits (e.g., Ireland, Poland, Spain), but some increase in others (e.g., Italy, Latvia, Lithuania). Although these reductions appear substantial in individual cases, most analysts agree that they do not represent a major barrier to the cost-effective use of JI/CDM. However, the EU-ETS does not accept credits from land use, land-use change and forestry (LULUCF) projects.

For Phase 3, the EU maintains its ban on using LULUCF credits within the ETS. However, it will permit up to 50% of the required reductions mandated under Phase 3 to be achieved through CDM or JI credits. For existing installations, this represents a total of 1.6 billion credits over the eight-year compliance period. Limits on use of Kyoto credits will be based on a facility's 2008- 2012 allocation (for an existing facilities) or its verified emissions during Phase 3 (for a new entrant or sector). The EC estimates that the minimum amount of Kyoto credits an existing facility will be able to use to comply with Phase 3 will be 11% of its 2008-2012 allocation, while new entrants and sectors will be able to use a minimum of 4.5% of their verified emissions during 2013-2020 (article 11a(8)). The precise percentages will be determined later.

Another flexibility mechanism, banking, is extended by the Directive from Phase 2 to Phase 3 in order to prevent a Phase 1 style collapse of allowance prices when the ETS transitions into Phase 3. In addition, the EU hopes that extending the trading period from five years to eight years, along with the steady, linear emissions reduction schedule, will increase certainty and stability in the allowance markets.

Phase 3 will introduce two other mechanisms designed to address price volatility. First, the EC is required under the Directive to examine whether the market for emission allowances is sufficiently protected from insider dealing or market manipulation. If not, the EC is to present proposals to ensure such protection to the EP and the Council (article 12(a)).

Second, the Directive provides that if the allowance price is more than three times the preceding two-year average for more than six consecutive months and the price is not based on market fundamentals, one of two measures may be taken. The first would allow Member States to shift forward the auctioning of some of its auctionable allowances. The second would allow Member States to auction up to 25% of the remaining allowances in the new entrants reserve (article 29(a)).

Expanding Coverage

Despite the EC's interest in expanding the ETS, its coverage in terms of industries included for Phase 2 is essentially the same as for Phase 1. The exception is for aviation. In December, 2006, the EC proposed bringing greenhouse gas emissions from civil aviation into the ETS in two phases.[50] As agreed to by the European Parliament in July 2008, all intra-EU and international flights will be included under the ETS beginning in 2012. Emissions would be capped at 97% of average 2004-2006 emissions with 85% of the allowances being allocated free to operators. The cap would be reduced to 95% in 2013. The cap and auctioning of allowances would be reviewed as a part of Phase 3 implementation.

Annex I of the Directive identifies three CO_2 emitting sectors for inclusion under the ETS: petrochemicals, ammonia, and aluminum. The ETS will also expand beyond CO_2 to include nitrous oxide (N_2O) emissions from nitric, adipic, and glyoxalic acid production, and perofluorocarbon (PFC) emissions from the aluminum sector. This would expand ETS covered emissions by 4.6% over Phase 2 allowance allocations, or about 100 million metric tons.[51] The harmonization and codification of eligibility criteria for combustion installations is expected to increase the coverage by a further 40-50 million metric tons.

To improve the cost-effectiveness of the ETS and reduce administrative costs, the Directive provides that small installations may be subject to other control regimes (such as carbon taxes) rather than included under the EU-ETS. Currently, the smallest 1,400 (10% of total installations covered) installations emit only 0.14% of total emissions covered. The Directive provides that Member States may opt to exclude installations that emit less than 25,000 metric tons annually from the EU-ETS (paragraph 11).

SUMMARY AND CONSIDERATIONS FOR U.S. CAP-AND-TRADE PROPOSALS

The United States is not a party to the Kyoto Protocol and no legislative proposal before the Congress would impose as stringent or rapid an emission reduction regime on the United States as Kyoto would have. Likewise, U.S. proposals to reduce emissions through 2020 are not as stringent as that provided in the EU Directive. However, through five years of carbon emissions trading, the EU has gained valuable experience. This experience, along with the process of developing Phase 3, may provide some insight into current cap-and-trade design issues in the United States.

Emission Inventories and Target Setting

The ETS experience with market trading and target setting confirms once again the central importance of a credible emissions inventory to a functioning cap-and-trade program.[52] The lack of credible EU-wide data on emissions was a direct cause of the ETS Phase 1 allowance market collapse in 2006. Arguably, the most important result of Phase 1 was the development of a credible inventory on which to base future targets and allocations.

In the United States, Section 821 of the 1990 Clean Air Act Amendments requires electric generating facilities affected by the acid rain provisions of Title IV to monitor carbon dioxide in accordance with EPA regulations.[53] This provision was enacted for the stated purpose of establishing a national carbon dioxide monitoring system.[54] As promulgated by EPA, regulations permit owners and operators of affected facilities to monitor their carbon dioxide emissions through either continuous emission monitoring (CEM) or fuel analysis.[55] The CEM regulations for carbon dioxide are similar to those for the acid rain program's sulfur dioxide CEM regulations. Those choosing fuel analysis must calculate mass emissions on a daily, quarterly, and annual basis, based on amounts and types of fuel used. As suggested by the EU-ETS experience, expanding equivalent data requirements to all facilities covered under a cap-and-trade program would be the foundation for developing allocation systems, reduction targets, and enforcement provisions.

Coverage

Despite economic analysis to the contrary, the EU decided to restrict Phase 1 ETS coverage to six sectors that represented about 40%-45% of the EU's CO_2 emissions.[56] This restriction was estimated to raise the cost of complying with Kyoto from 6 billion euro annually to 6.9 billion euro (1999 euro) compared with a comprehensive trading program. A variety of practical, political, and scientific reasons were given by the EC for the decision.[57]

The experience of the ETS up to now suggests that adding new sectors to an existing trading program is a difficult process. As noted above, a stated goal of the EC is to expand the coverage of the ETS. However, the experience of Phase 1 did not result in the addition of any new sector until the last year of Phase 2 when aviation will be included. The EU will expand its coverage with Phase 3, but the ETS will still cover fewer sectors emitting greenhouse gases than provided under most U.S. proposals.

U.S. cap-and-trade proposals generally fall into one of two categories. Most bills are more comprehensive than the ETS, covering 80% to 100% of the country's greenhouse gas emissions. At a minimum, they include the electric utility, transportation, and industrial sectors; disagreement among the bills center on the agricultural sector and smaller commercial and residential sources. In some cases discretion is provided EPA to exempt sources if serious data, economic, or other considerations dictate such a resolution.

A second category of bills focuses on the electric utility industry, representing about 33% of U.S. greenhouse gases and therefore less comprehensive than the ETS. Sometimes including additional controls on non-greenhouse gas pollutants, such as mercury, these bills focus on the sources with the most experience with emission trading and the best emissions data. Other sources could be added as circumstances dictate.

As noted, the EU's experience with the ETS suggests that adding sectors to an emission trading scheme can be a slow and contentious process. If one believes that the electric utility sector is a cost-effective place to start addressing greenhouse gas emissions and that there is sufficient time to do the necessary groundwork to eventually add other sectors, then a phased-in approach may be reasonable. If one believes that the economy as a whole needs to begin adjusting to a carbon- constrained environment to meet long term goals, then a more comprehensive approach may be justified. The ETS experience suggests the process doesn't necessarily get any easier if you wait.

Allocation Schemes

Setting up a tradeable allowance system is a lot like setting up a new currency.[58] Allocating allowances is essentially allocating money with the marketplace determining the exchange rate. As noted above, the free allocation scheme used in the ETS has resulted in "windfall profits" being received by allowance recipients. As stated quite forcefully by Deutsche Bank Research:

> The most striking market outcome of emissions trading to date has been the power industry's windfall profits, which have sparked controversy. We are all familiar with the background: emissions allowances were handed out free of charge to those plant operators participating in the emissions trading scheme. Nevertheless, in particular the producers of electricity succeeded in marking up the market price of electricity to include the opportunity-cost value of the allowances. This is correct from an accounting point of view, since the allowances do have a value and could otherwise be sold. Moreover, emissions trading cannot work without price signals.[59]

The free allocation of allowances in Phase 1 and 2 of the ETS incorporates two other mechanisms that create perverse incentives and significant distortions in the emissions markets: new entrant reserves and closure policy. Combined with an uncoordinated and spotty benchmarking approach for both new and existing sources, the result is a greenhouse gas reduction scheme that is influenced as much or more by national policy than by the emissions marketplace.

The expansion of auctions for Phase 3 of the ETS could simplify allocations and permit market forces to influence compliance strategies more fully. Most countries did not employ auctions at all during Phase 1 and auctions continue to be limited under Phase 2. No country combined an auction with a reserve price to encourage development of new technology. The EC limited the amount of auctioned allowances to 10% in Phase 2: a limit no country chose to meet. Efforts to expand auctions met opposition from industry groups, but attracted support from environmental groups and economists. The Phase 3 increased use of auctioning through 2020 will represent a major development for the scheme.

Currently, all U.S. cap-and-trade proposals have some provisions for auctions, although the amount involved is sometimes left to EPA discretion. Most specify a schedule that provides increasing use of auctions from 2012 through the mid-2030s with a final target of 66%-100% of total allowances auctioned. Funds would be used for a variety of purposes, including programs to encourage new technologies. Some proposals include a reserve price on some auctions to create a price floor for new technology.

Like the situation in the ETS, most U.S. industry groups either oppose auctions outright or want them to be supplemental to a base free allocation. Given the experience with the ETS where the EC and individual governments have been unwilling or unable to move away from free allocation, the Congress, like the EU, may ultimately be asked to consider specifying any auction requirement if it wishes to incorporate market economics more fully into compliance decisions.

Flexibility and Price Volatility

Despite EU rhetoric during the Kyoto Protocol negotiations, it moved into Phase 2 without a significant restriction on the use of CDM and JI credits. This embracing of project credits will significantly increase the flexibility facilities have in meeting their reduction targets. In addition, Phase 2 includes the use of banking to increase flexibility across time by allowing banked allowances to be used in Phase 3. Each of these market mechanisms is projected to reduce both the EU's Kyoto compliance costs and allowance price volatility. These flexibility mechanisms will be extended into Phase 3 with modifications.

Unfortunately, Phase 1 experience with the ETS did not provide much useful information on the value of market mechanisms or financial instruments in reducing costs or price volatility. The combination of poor emissions inventories, non-use of project credits, and time-limited allowances with effectively no banking resulted in extreme price volatility in Spring 2006, and virtually worthless allowances by mid-2007. The real test for the mechanisms employed by the ETS to create a stable allowance market is Phase 2. Initial indications are that a mature market for allowances appears to be developing, although, like most commodities markets, the allowance market can still be volatile at times.

Phase 3 is introducing two new mechanisms in the ETS to further respond to volatility not based on market fundamentals. However, the actual effectiveness of these mechanisms is yet to be proven.

Like the ETS, U.S. cap-and-trade proposals would employ a combination of devices to create a stable allowance market and encourage flexible, cost-effective compliance strategies by participating entities. All include banking. All include use of offsets, although some would place substantial restrictions on their use. Some proposals have incorporated a "safety valve" that would effectively place a ceiling on allowance prices, while others would create a Carbon Market Efficiency Board to observe the allowance market and implement cost-relief measures if necessary. Finally, some incorporate strategic reserves auctions, similar in concept to the EU forward auctioning mechanism, to increase allowance supply without busting the emission cap. Some see this as a more flexible response with the potential for avoiding or mitigating the environmental impacts of a safety valve (i.e., increased emissions).

Additionally, concern has been expressed in the United States about the regulation of allowance markets and instruments. Based on experience with the ETS, the potential for speculation and manipulation could extend beyond the emission markets. Analysis of ETS allowance prices during Phase 1 suggests the most important variables in determining allowance price changes were oil and natural gas price changes.[60] This apparent linkage between allowance price changes and price changes in two commodities markets raises the possibility of market manipulation, particularly with the inclusion of financial instruments such as options and futures contracts. The concern is sufficient for the Directive to require the EC to examine the situation and the current protections against such activities. Congress may ultimately be asked to consider whether the Securities and Exchange Commission, Federal Energy Regulatory Commission, the Commodities Futures Trading Commission, or other body should have enhanced regulatory and oversight authority over such instruments.[61]

End Notes

[1] Readers unfamiliar with the workings of the European Union may want to read CRS Report RS21372, *The European Union: Questions and Answers*, by Kristin Archick and Derek E. Mix.

[2] Six gases are included under the Kyoto Protocol: carbon dioxide, methane, nitrous oxide, hydrofluorocarbons, perfluorocarbons, and sulfur hexafluoride. The United States has not ratified the Kyoto Protocol and, therefore, is not covered by its provisions. For more information on the Kyoto Protocol, see CRS Report RL33826, *Climate Change: The Kyoto Protocol, Bali "Action Plan," and International Actions*, by Jane A. Leggett.

[3] Norway, a non-EU country, also has instituted a CO_2 trading system which is currently linked with the EU-ETS. Various other countries and a state-sponsored regional initiative located in the northeastern United States involving several states are developing mandatory cap-and-trade system programs, but are not operating at the current time. For a review of these emerging programs, along with other voluntary efforts, see CRS Report RL33812, *Climate Change: Action by States to Address Greenhouse Gas Emissions*, by Jonathan L. Ramseur.

[4] P.L. 101-549, Title IV (November 15, 1990).

[5] For further background on the ETS, see CRS Report RL34150, *Climate Change and the EU Emissions Trading Scheme (ETS): Kyoto and Beyond*, by Larry Parker.

[6] More information, including relevant directives, on the EU-ETS is available on the European Union's website at http ://europa. eu.int/scadplus/leg/en/lvb/l28012.htm.

[7] Directive 2003/87/EC of the European Parliament and of the Council of 13 October 2003 establishing a scheme for greenhouse gas emissions allowance trading within the Community and amending Council Directive 96/61/EC.

[8] Commission of the European Communities, *Communication from the Commission: Progress towards Achieving the Kyoto Objectives* (November 19, 2008). Other reductions are to be achieve through regulatory measures, such as a CO_2 emissions standard for automobiles.

[9] A. Denny Ellerman and Barbara K. Buchner, "The European Union Emissions Trading Scheme: Origins, Allocations, and Early Results," *Environmental Economics and Policy* (Winter 2007), pp. 69-70; and International Emissions Trading Association, "IETA Position Paper on EU ETS Marking Functioning," (no date), p. 3.

[10] For more on the effect of the ETS on Kyoto mechanisms, see A. Denny Ellerman and Barbara K. Buchner, "The European Union Emissions Trading Scheme: Origins, Allocations, and Early Results," *Environmental Economics and Policy* (Winter 2007), p. 84; and International Emissions Trading Association, "IETA Position Paper on EU ETS Market Functioning" (no date), p. 2. For more information on the Kyoto Protocol mechanisms, see CRS Report RL33826, *Climate Change: The Kyoto Protocol, Bali "Action Plan," and International Actions*, by Jane A. Leggett.

[11] A Denny Ellerman and Barbara K. Buchner, "The European Union Emissions Trading Scheme: Origins, Allocations, and Early Results," *Environmental Economics and Policy* (Winter 2007), p. 84.

[12] European Commission, Directorate General for Environment, *Review of EU Emissions Trading Scheme: Survey Highlights*, (November 2005), pp. 5-7.

[13] For a further discussion of Phase 2 implementation issues, see CRS Report RL34 150, *Climate Change and the EU Emissions Trading Scheme (ETS): Kyoto and Beyond*, by Larry Parker.

[14] For more information, see CRS Report RL33970, *Greenhouse Gas Emission Drivers: Population, Economic Development and Growth, and Energy Use*, by John Blodgett and Larry Parker.

[15] On the role of modeling in the first phase, see A Denny Ellerman and Barbara K. Buchner, "The European Union Emissions Trading Scheme: Origins, Allocations, and Early Results," 1 *Environmental Economics and Policy* 1 (Winter 2007), pp. 72-73.

[16] Regina Betz and Misato Sato, "Emissions Trading: Lessons Learnt from the 1st Phase of the EU ETS and Prospects for the 2nd Phase," 6 *Climate Policy* (2006), p. 354.

[17] For a further discussion, see CRS Report RL33581, *Climate Change: The European Union's Emissions Trading System (EU-ETS)*, by Larry Parker.

[18] A Denny Ellerman and Barbara K. Buchner, "The European Union Emissions Trading Scheme: Origins, Allocations, and Early Results," 1 *Environmental Economics and Policy* 1 (Winter 2007), p. 83.

[19] European Commission, *Communication from the Commission to the Council and to the European Parliament on the assessment of national allocation plans for the allocation of greenhouse gas emission allowances in the second period of the EU Emissions Trading Scheme,* COM(2006) 725 final (November 29, 2006), p. 11.

[20] For a further discussion, see Joseph Kruger, Wallace E. Oates, and William A. Pizer, "Decentralization in the EU Emissions Trading Scheme and Lessons for Global Policy, 1 *Environmental Economics and Policy* 1 (Winter 2007), p. 126; and, Frank J. Convery and Luke Redmond, "Market and Price Development in the European Union Emissions Trading Scheme, 1 *Environmental Economics and Policy* 1 (Winter 2007), pp. 96-7, 107.

[21] International Emissions Trading Association, "IETA Position Paper on EU ETS Market Functioning," (no date), p. 2.

[22] Once the EC has approved a country's NAP, including the total number of allowances and the allocation to each covered entity, the allocations can not be re-visited. Attempts to include provisions permitting such post-approval adjustments to a facility's allocation have been uniformly rejected by the EC.

[23] European Commission, *Communication from the Commission to the Council and to the European Parliament on the assessment of national allocation plans for the allocation of greenhouse gas emission allowances in the second period of the EU Emissions Trading Scheme,* COM(2006) 725 final (November 29, 2006), p 8; and, A Denny Ellerman and Barbara K. Buchner, "The European Union Emissions Trading Scheme: Origins, Allocations, and Early Results," 1 *Environmental Economics and Policy* 1 (Winter 2007), p. 71.

[24] On the mixed record of the EU-ETS and the need for allowance scarcity to a functioning emissions market, see Eric Haymann, *EU Emission Trading: Allocation Battles Intensifying*, Deutsche Bank Research (March 6, 2007). For a generally positive view of ETS market development, see Frank J. Convery and Luke Redmond, "Market and Price Development in the European Union Emissions Trading Scheme, 1 *Environmental Economics and Policy* 1 (Winter 2007), pp. 97-106. For a more negative view, see Karsten Neuhoff, Federico Ferrario, Michael Grubb, Etienne Gabel, and Kim Keats, "Emissions Projections 2008-2012 Versus NAPs II," 6 *Climate Policy* 5 (2006), pp. 395-410.

[25] European Environment Agency, *Greenhouse Gas Emissions Trends and Projections in Europe 2009*, (2009) p. 11.

[26] Cyprus and Malta are not Annex 1 countries.

[27] European Environmental Agency, *Greenhouse Gas Emission Trends and Projections in Europe 2009* (2009), p. 12.

[28] European Commission, Directive 2009/29/EC *of the European Parliament and of the Council of 23 April 2009 amending Directive 2003/87/EC so as to improve and extend the greenhouse gas emission allowance trading system of the Community* (Brussels, April 23, 2009). Hereinafter referred to as the Directive.

[29] European Commission, *Limiting Global Change to 2 degrees Celsius: The Way Ahead for 2020 and Beyond* (Brussels, January 10, 2007), p. 23.

[30] Joachim Schleich, Regina Betz, and Karoline Rogge, *EU Emissions Trading—Better Job Second Time Around?* Fraunhofer Institute System and Innovation Research (February 2007), p. 23.

[31] Commission of the European Communities, Directive 2003/87/EC, available at *http://eur-lex.europa.eu/LexUriServ/ LexUriServ.do*?uri=OJ:L:2003 :275:0032:0046:EN:PDF.

[32] European Commission, *Proposal for a Directive of the European Parliament and of the Council amending Directive 2003/87/EC so as to improve and extend the greenhouse gas emission allowance trading system of the Community* (Brussels, January 23, 2008).

[33] European Commission, *Directive 2009/29/EC of the European Parliament and of the Council amending Directive 2003/8 7/EC so as to improve and extend the greenhouse gas emission allowance trading system of the Community* (Brussels, April 23, 2009).

[34] For a further discussion of auctioning and the ETS, see Cameron Hepburn, *et. al*., "Auctioning of EU ETS phase II allowances: how and why?" 6 *Climate Policy* (2006), pp. 137-160.

[35] A Denny Ellerman and Barbara K. Buchner, "The European Union Emissions Trading Scheme: Origins, Allocations, and Early Results," 1 *Environmental Economics and Policy* 1 (Winter 2007), p. 73.

[36] Joseph Kruger, Wallace E. Oates, and William A. Pizer, "Decentralization in the EU Emissions Trading Scheme and Lessons for Global Policy," 1 *Environmental Economics and Policy* 1 (Winter 2007), p. 114.

[37] E.g., Deutsche Bank Research, *EU Emission Trading: Allocation Battles Intensifying*, (March 6, 2007) pp. 2-3; and Regina Betz and Misato Sato, "Emissions Trading: Lessons Learnt from the 1st Phase of the EU ETS and Prospects for the 2nd Phase," 6 *Climate Policy* (2006), p. 353.

[38] A Denny Ellerman and Barbara K. Buchner, "The European Union Emissions Trading Scheme: Origins, Allocations, and Early Results," 1 *Environmental Economics and Policy* 1 (Winter 2007), p. 85.

[39] For more information, see CRS Report RL34 150, *Climate Change and the EU Emissions Trading Scheme (ETS): Kyoto and Beyond*, by Larry Parker.

[40] Martina Priebe, *Distributional Effect of Carbon-allowance Trading* (Cambridge, January 12, 2007). Also, see Eurochambres, *Review of the EU Emission Trading System* (June 2007), p. 5.

[41] See *Position of the European Parliament adopted at the first reading on 17 December 2008 with a view to the adoption of Directive 2009/.../EC of the European Parliament and of the Council amending Directive 2003/8 7/EC so as to improve and extend the greenhouse gas emission allowance trading system of the Community* (December 17, 2008).

[42] For example, the U.S. acid rain program provides no allocation of allowances to new entrants; instead, an EPA sanctioned auction is held annually to ensure that allowances are available to new entrants. New entrants can also obtain allowances from existing sources willing to sell them, either directly, through the EPA auction, or via a broker.

[43] A Denny Ellerman and Barbara K. Buchner, "The European Union Emissions Trading Scheme: Origins, Allocations, and Early Results," 1 *Environmental Economics and Policy* 1 (Winter 2007), p. 75.

[44] Karoline Rogge, Joachim Schleich, and Regina Betz, *An Early Assessment of National Allocation Plans for Phase 2 of EU Emission Trading*, Fraunhofer Institute System and Innovation Research (January 2006).

[45] For more information on climate change and competitiveness issues, see CRS Report R40914, *Climate Change: EU and Proposed U.S. Approaches to Carbon Leakage and WTO Implications*, by Larry Parker and Jeanne J. Grimmett.

[46] Julia Reinaud, *Issues Behind Competitiveness and Carbon Leakage: Focus on Heavy Industry* (October 2008), p. 6.

[47] Carbon Trust, *EU ETS Impacts on Profitability and Trade* (January 2008), p. 4; and The Climate Group, *The Effects of EU Climate Legislation on Business Competitiveness; A Survey and Analysis* (September 2009), p. 8.

[48] European Commission, *Commission Decision of 24 December 2009 determining, pursuant to Directive 2003/87/EC of the European Parliament and of the Council, a list of sectors and subsectors which are deemed to be exposed to a significant risk of carbon leakage* (Brussels, 2009).

[49] European Commission, *Draft Commission Decision of 18 September2009 determining, pursuant to Directive 2003/87/EC of the European Parliament and of the Council, a list of sectors and subsectors which are deemed to be exposed to a significant risk of carbon leakage* (Brussels, 2009).

[50] European Commission, *Proposal for a Directive of the European Parliament and of the Council amending Directive 2003/8 7/EC so as to include aviation activities in the scheme for greenhouse gas emission allowance trading within the Community* (Brussels, December 12, 2006).

[51] European Commission, *Proposal for a Directive of the European Parliament and of the Council amending Directive 2003/87/EC so as to improve and extend the greenhouse gas emission allowance trading system of the Community* (Brussels, January 23, 2008), p. 4.

[52] As stated by CRS in 1992: "For an economic incentive system to be effective, several preconditions are necessary. Perhaps the most important is data about the emissions being controlled. Such data are important to levy any tax, allocate any permits, and enforce any limit." CRS Issue Brief IB92125, *Global Climate: Proposed Economic Mechanisms for Reducing* CO_2, by Larry Parker (archived November 16, 1994), p. 9.

[53] Section 821, *1990 Clean Air Act Amendments* (P.L. 101-549, 42 USC 765 1k).

[54] S.Rept. 101-952.

[55] See 40 CFR 75.13, along with appendix G (for CEMs specifications) and appendix F (for fuel analysis specifications.

[56] For more background, see CRS Report RL3358 1, *Climate Change: The European Union's Emissions Trading System (EU-ETS)*, by Larry Parker.

[57] Ibid., p 3.

[58] Unlike a carbon tax which uses the existing currency system to control emissions—be it euro or dollars.

[59] Deutsche Bank Research, *EU Emission Trading: Allocation Battles Intensifying* (March 6, 2007), p. 2.

[60] Maria Mansanet-Bataller, Angel Pardo, and Enric Valor, "CO_2 Prices, Energy and Weather," 28 *The Energy Journal* 3 (2007), pp. 73-92.

[61] For a discussion of regulation of allowances as a commodity and implications for a greenhouse gas emissions market, see CRS Report RL34488, *Regulating a Carbon Market: Issues Raised By the European Carbon and U.S. Sulfur Dioxide Allowance Markets*, by Mark Jickling and Larry Parker.

In: The Role Emissions in Climate Change Policy ISBN: 978-1-61728-986-6
Editors: Heather C. Perkins

Chapter 3

REGULATING A CARBON MARKET: ISSUES RAISED BY THE EUROPEAN CARBON AND U.S. SULFUR DIOXIDE ALLOWANCE MARKETS*

Mark Jickling and Larry Parker

SUMMARY

Both the European Union's Emissions Trading Scheme (EU-ETS) and the U.S. Title IV sulfur dioxide (SO_2) program provide insights into regulatory issues that may face any future U.S. carbon market. From the initial operations of the EU-ETS, the 2006 price crash raised questions about the adequacy of market regulation. In particular, some suspect that information about allocations leaked before official publication, and that certain traders profited from this knowledge.

Title IV's longer trading history reveals two important trends: (1) an increasing trend toward diverse and non-traditional participants that is likely to continue under a carbon market; (2), an increasing use of financial instruments to manage allowance price risk that is likely to expand under a carbon market as a hedge against price uncertainty. Indeed, a carbon market may look more

* This is an edited, reformatted and augmented edition of a United States Congressional Research Service publication, Report RL34488, dated February 26, 2010.

like other energy markets, such as natural gas and oil, than the somewhat sedate SO_2 allowance market.

Regulation of emissions trading would have to consider two kinds of fraud and manipulation: fraud by traders or intermediaries against other investors, and sustained price manipulation. Four agencies could have roles in the regulation of an emissions market, each with its own attributes that may contribute to effective regulation.

The Commodities Futures Trading Commission (CFTC) currently oversees the Title IV program and its mission most closely resembles what a regulator of a future carbon market would do, including market surveillance to prevent or detect fraud and manipulation. The major weakness of the CFTC, according to some, is that it lacks resources and the statutory mandate to do its job. Current derivatives reform proposals would greatly enlarge its regulatory scope.

The Securities and Exchange Commission (SEC) is much larger than the CFTC, but it also faces resource and capability issues. While the CO_2 market will resemble commodities markets more closely than securities, SEC has some appropriate regulatory tools applicable to an emissions market.

The Environmental Protection Agency (EPA) would likely be responsible for the primary market in allowances. However, EPA lacks experience comparable to that of the CFTC and SEC in regulating trading markets, although the data it gathered in the primary market could be critical to oversight of the secondary market.

Federal Energy Regulatory Commission (FERC) was granted oversight authority over bulk electricity and interstate natural gas markets in 2005. Its experience with market surveillance and enforcement is thus limited in comparison to the SEC and CFTC, and it does not play an active role in overseeing the Title IV market.

It is possible that no single regulator would have clear jurisdiction, as is the case in the Title IV program. This kind of regulatory fragmentation has not always worked well. An umbrella group to monitor markets and provide a forum for regulatory coordination might help to prevent regulatory gaps or conflicts in the market.

INTRODUCTION

A number of congressional proposals to advance programs that reduce greenhouse gases have been introduced in the 111th Congress. Proposals receiving particular attention would create market-based greenhouse gas reduction programs along the lines of the trading provisions of the current sulfur dioxide (acid rain) reduction program established by Title IV of the 1990 Clean Air Act Amendments.[1] These "cap-and-trade" schemes would impose a ceiling (cap) on total annual emissions of greenhouse gases and establish a market in pollution rights, called allowances, between affected entities. An allowance would be a limited authorization by the government to emit one metric ton of carbon dioxide equivalent (CO_2e), and could be bought and sold (traded) or held (banked) by participating parties.

These domestic proposals have parallels with the programs being implemented in Europe to meet its obligations under the Kyoto Protocol and the Copenhagen Accord. Specifically, the European Union (EU) has decided to implement a cap-and-trade program, along with other market-oriented mechanisms permitted under the Kyoto Protocol, to help it achieve compliance at least cost.[2] The EU's decision to use emission trading to implement the Kyoto Protocol is at least partly based on the successful emissions trading program used by the United States to implement its acid rain control program.

These two operating cap-and-trade programs—the U.S.'s acid rain program and the EU's climate change program—may provide insights for the design of a domestic greenhouse gas reduction scheme.[3] However, while the experiences of the EU system directly relate to the greenhouse gas reduction initiative of the domestic legislative proposals, it has operated only five years (see text box). The acid rain control program has a longer operating history, although the control scheme differs in some important ways—e.g., it is internal to one nation and involves fewer types of sources.

Among the lessons that Phase 1 of the European Trading System may have for a similar U.S. program is that allowance prices are linked to the price of other energy commodities.[4] Analysis of ETS allowance prices during Phase 1 suggests the most important variables in determining allowance price changes have been oil and natural gas price changes.[5] For example, when natural gas, the cleaner fuel, becomes more expensive relative to oil, industrial users may switch to oil, creating increased demand for allowances. This suggests that traders will pursue arbitrage strategies involving simultaneous transactions in allowances and oil and gas contracts. For example, a trader anticipating a rise in the price of oil might take a position in allowances in the expectation that the two prices would move in tandem. Since there is widespread suspicion that excessive speculation by hedge funds and others has affected energy prices in recent years,[6] the possibility that the price of allowances could also be subject to distortion or manipulation will be a policy concern.

> The EU's Emissions Trading System (ETS) covers more than 10,000 energy intensive facilities across the 27 EU Member countries, including oil refineries, powerplants over 20 megawatts (MW) in capacity, coke ovens, and iron and steel plants, along with cement, glass, lime, brick, ceramics, and pulp and paper installations. Covered entities emit about 45% of the EU's carbon dioxide emissions. The trading program covers neither CO2 emissions from the transportation sector, which account for about 25% of the EU's total greenhouse gas emissions, nor emissions of non-CO2 greenhouse gases, which account for about 20% of the EU's total greenhouse gas emissions. A "Phase 1" trading period ran from 2005 through 2007. A second, Phase 2, trading period began in 2008, covering the period of the Kyoto Protocol, with a third one finalized to begin for 2013. (For further background on the ETS, see CRS Report RL34 150, *Climate Change and the EU Emissions Trading Scheme (ETS): Kyoto and Beyond*, by Larry Parker, and CRS Report R41049, Climate Change and the EU Emissions Trading Scheme (ETS): Looking to 2020, by Larry Parker Relevant directives on the EU-ETS are available at http://ec. europa.eu/environment/climat/ emission.htm#brochure.)

This chapter examines the ETS and Title IV sulfur dioxide cap-and-trade program, with a focus on the market activity and the current regulatory overlay. From that discussion, observations are drawn about implications for regulating a future greenhouse gas trading scheme in the United States. No current U.S. cap-and-trade proposal has specific provisions with respect to carbon allowance financial instruments or who would regulate such a market or its participants.

THE EUROPEAN EMISSIONS TRADING SYSTEM (ETS)

Background

The European Emissions Trading System (ETS) is by far the largest market for greenhouse gas emissions allowances. Trading began in 2005, when allowances were issued by the 27 member states of the European Union to over 10,000 electric utility and industrial sources of CO_2. In 2008, allowances for 2.7 billion metric tons of carbon emissions changed hands, with a financial value of 67 billion euro.[7]

Secondary market trading involves not only companies to whom allowances are originally allocated, but also a range of brokers and intermediaries. Trading occurs in the form of bilateral agreements and on over-the-counter (OTC) markets, both of which are essentially unregulated (other than an obligation to report all transactions to the national registry). Trading also takes place on financial exchanges, which are subject to various forms and degrees of regulation. The leading exchange market is the European Climate Exchange (ECX) in London, which handles about 75% of exchange-traded volume. The other exchanges with significant trading volumes are BlueNext (formerly Powernext), based in Paris, with about a 14% market share; Nord Pool, a Scandinavian electrical power exchange, which handles about 8%; and the European Energy Exchange (EEX), in Leipzig, Germany, with about 4% of volume.[8]

The basic unit of trading is the European Union Allowance (EUA), which permits the holder to emit one metric ton of CO_2. Allowances themselves are bought and sold in the spot market, but there are several additional forms of emissions trading:

- *Term contracts* call for a specified number of deliveries of allowances to take place over a period of time.
- *Forward contracts* are sales where delivery is to be made at a future date, but at today's price.
- *Futures contracts,* which are traded only on exchanges, give holders the right to buy or sell allowances over the term of the contract at the price that prevailed when the contract was made. They gain or lose value as the market price of the underlying allowance fluctuates, and they allow financial speculators who do not deal in physical CO_2 themselves to participate in the market.

- *Swaps* are economically equivalent to futures, but are traded over-the-counter rather than on an exchange. The value of a swap is linked to the price of the underlying allowance. The two counterparties may structure the contract any way they like, but the essence is that one will pay the other if the price rises, and vice versa.[9]
- *Spreads* are two or more transactions that take place simultaneously. For instance, a firm with a surplus of 2008 EUAs that anticipates a shortage in 2009 might sell the former and buy the latter. The price would be expressed as the differential between the two contract prices at the time the trade was made.

In addition to the market in EUAs, there is a related market in certified emission reductions (CERs). A CER represents a reduction of CO_2 emissions by one ton outside the EU, generally in the developing world. CERs are created under the United Nations Kyoto Protocol. The United Nations, through its Clean Development Mechanism, verifies that the reductions have in fact taken place, and issues the corresponding CERs, which may be submitted by EU CO2 emitters in place of EUAs within certain limits.[10] To EU market participants, CERs and EUAs are close substitutes. Trading in CERs takes the same forms: regulated or unregulated spot markets and forwards, futures, and other derivative instruments.

What Is Regulated

Implementation of the ETS proceeded without a formal directive or specification as to secondary trading venues or regulation:

> The legal framework of the ETS does not lay down how and where trading in allowances should take place. Companies and other participants in the market may trade directly with each other or buy and sell via a broker, exchange, or any type of market intermediary that may spring up to take advantage of a new market of significant size.[11]

The decision was to let trading be shaped by market forces. As a result, the act of buying and selling allowances does not in itself subject one to regulation, or make one a regulated entity. This is not to say, however, that the EU envisioned a completely unregulated market. Instead, allowance traders come under regulation when they engage in financial practices that are regulated regardless of the nature of the underlying interest or instrument. Similarly, when transactions take place on regulated markets, emissions trading is regulated in the same way as other transactions executed on those markets. The degree of regulation depends on two things: (1) does the instrument being traded meet the definition of a financial instrument or product, and (2) is the trade taking place on a regulated market? Although financial regulation has been harmonized in the EU to a significant degree, there remains some variation in these definitions and in market regulation among EU member states.

On the first key point—what is being traded—there appears to be general agreement among national laws. The simple sale of an allowance by one party to another does not subject either party to regulation.[12] A firm may sell its EUAs the way it sells any other piece of property.

If, however, what is being transferred is not an EUA itself, but rather a financial contract that conveys rights to an EUA, or that is based on future delivery or the future price of an allowance, the trade meets the definition of a financial transaction or instrument in many countries and may be regulated. A person arranging or participating in such a transaction may come under the jurisdiction of banking or securities laws and be required to register as an intermediary or broker.[13] Certain forms of trades, like futures contracts, may be restricted to regulated futures exchanges.

Trades that are otherwise unregulated may come under government oversight if they occur on a regulated market. Several of the European emissions exchanges provide a platform for spot market trades, as well as financial contracts. These spot trades are subject to the same types of market surveillance as the financial contracts (some of which could not be legally executed off- exchange), and the same registration and reporting requirements apply. In addition, if an unregulated over-the-counter (OTC) trade, such as a swap or forward contract, is processed and guaranteed by a regulated clearing house, that trade is also open to the scrutiny of financial regulators.[14]

How Exchanges Are Regulated

National regulations vary within the EU. The differences, however, continue to diminish as a result of both EU harmonization directives and the globalization of markets, which is most visible in the numerous cross-border mergers that continue to occur among securities and futures exchanges. Many of the differences that remain are matters of terminology rather than substance.

In general, the exchanges where emissions and emissions derivatives are traded are subject to a regulatory scheme broadly comparable to the regulation of the futures exchanges in the United States by the Commodity Futures Trading Commission (CFTC).[15] The exchange itself must satisfy numerous conditions of registration, which include market surveillance to deter fraud and manipulation, various reporting requirements including publication of trade and price data, and so on. Exchange rules are generally subject to regulatory oversight or approval.

Access to the trading mechanisms is generally limited to exchange members; others must trade as customers of a member intermediary. Exchange members may be required to register with a regulatory agency and in all cases are subject to exchange rules and financial standards. (Apart from their exchange memberships, brokerage firms may also be subject to registration and regulation by financial authorities because of the nature of their business.) Some features of the regulatory environments of the four largest EU emissions exchanges are set out in an appendix to this chapter.

Clearing houses may be part of an exchange or stand-alone entities. They are also subject to registration requirements, generally administered by the exchange regulator. In addition, because they represent a concentration of financial risk, they may be subject to central bank safety and soundness regulation.

The fact that a given market organization may be international in scope may require it to report to multiple national regulators.[16] This does not appear to have been a barrier to the development of the market, or a source of market fragmentation. As noted above, London has become the trading center, with about three-quarters of all secondary derivatives trading. (In commodity markets, trading tends to gravitate to the single market providing most liquidity. Before the EU drive for a single financial market, this impulse was blocked by laws and regulations in several European countries intended to preserve a monopoly for local exchange and traders.)

Lessons from the ETS

The fact that the ETS was implemented without a formal, government-devised blueprint for secondary trading does not appear to have caused significant problems in the market. According to one study, "This apparent confidence in market ingenuity has proved well-founded."[17] Trading volumes are growing rapidly, suggesting that investors do not view the market as being rigged against them or subject to manipulation by insiders.

This does not mean that the first phase of the ETS (from 2005 through 2007) was an unqualified success. In fact, many regard it as a disaster. The major problem, however, did not come from the secondary market but rather from the primary market process of allocating and issuing EUAs.

When phase one began in 2005, trading commenced before the formal allocations had been made. Thus, prices were initially based on traders' expectations of the number of EUAs that would be issued.[18] Allocations were not made at the EU level, but were left to the discretion of individual countries. When the allocations were finally made, it became apparent that some countries had been very generous and that the total supply of EUAs was going to be much higher than the market anticipated. As a result, the market crashed in the spring of 2006: the price per EUA dropped in a few weeks from over €31 to less than €11 (and by the end of phase one, in 2007, fell below one euro).[19]

The price crash, by weakening the financial incentive to reduce pollution, undermined the basic rationale for the ETS. Although the problem did not originate in the secondary market, it did raise questions about the adequacy of market regulation. In particular, many observers suspect that information about the size of allocations leaked before official publication, and that certain traders profited from this knowledge. Traders with nonpublic information on certified emissions data—which gives an indication of future demand for allowances—may also have profited:

> Unauthorized leaks of verified emission data for 2005 in several countries created information asymmetries and undue opportunities for some businesses market participants. This has highlighted the need for strict rules and procedures for handling of price-sensitive information along the lines that is common in more mature financial markets.[20]

A number of theoretical studies have raised the possibility of price manipulation by a dominant firm or by a few large firms acting in collusion—while thousands of pollution sources receive EUAs, a few large power generators account for a disproportionate share.[21] Some observers have raised the possibility that such manipulation may have occurred during the price crash:

> The fact that knowledge about excess allowances has gradually become known without the market reacting immediately was seen by some analysts as a sign that there might have been collusion of the big power companies, which in essence are the major buyers and equally benefit by a high allowance price through higher power prices.[22]

It is likely, however, that any undue market power accruing to a few large consumers of allowances—in the EU or the United States—will diminish as emissions trading becomes global in scope, particularly if markets are linked electronically.

It should be noted that EU regulators have not brought any enforcement actions based on manipulation of emissions prices (in public, at least). On balance, the secondary market in emissions allowances that has evolved without central direction, and in a combination of regulated and unregulated venues, appears to have functioned well. For the most part, the incentives of market participants and regulators are in alignment: both want an efficient and transparent price discovery mechanism, and a financially-sound market free of manipulation and fraud.

This is not to say that the EU is not concerned about the possibility for manipulation and fraud in the carbon markets. In the final directive setting up Phase 3 of the EU-ETS, the European Commission (EC) is required to examine whether the market for emission allowances is sufficiently protected from insider dealing or market manipulation. If not, the EC is to present proposals that ensure such protection to the European Parliament and the Council of Ministers.[23]

THE U.S. SULFUR DIOXIDE TRADING PROGRAM (TITLE IV)

Background

Title IV of the 1990 Clean Air Act Amendments supplements the sulfur dioxide (SO_2) commandand-control system of the Clean Air Act (CAA) by limiting total SO_2 emissions from electric generating facilities to 8.95 million tons annually, beginning in the year 2000.[24] Title IV essentially caps SO_2 emissions at individual utility sources operating before enactment of the CAA in 1990 (known as "existing sources") through a tonnage limitation, and at those plants beginning operation after enactment (known as "new sources") through an emissions offset requirement. SO_2 emissions from most existing sources are capped at a specified emission rate times a historical average fuel consumption level. Beginning January 1, 2000, SO_2 emissions from new plants commencing operation after enactment must be offset—in effect, the emissions cap for new sources is zero. Their allowances come from emissions reductions at existing facilities. The program was implemented through a two-phase process with the final phase beginning in 2000.

To implement the SO_2 reduction program, the law creates a comprehensive permit and emissions allowance system (cap-and-trade program). An allowance is a limited authorization to emit a ton of SO_2 during or after a specified year. Issued by EPA, the allowances are allocated to existing power plant units in accordance with formulas delineated in the law. The owner of the facility receives the allowances for a given plant regardless of the actual operation of the plant. For example, an owner may choose to shut down an existing power plant and use those allowances to offset emissions from two newer, cleaner facilities. As noted, generally, a power plant that commences operation after enactment receives no allowances, requiring new units to obtain allowances from those with allowances, or purchase them at an EPA-sponsored auction, in order to operate after 2000. An owner may trade allowances nationally as well as bank allowances for future use or sale.

If an affected unit does not have sufficient allowances to cover its emissions for a given year, it is subject to an emission penalty of $2,000 (1990$, indexed to inflation) per ton of excess SO_2 , and it submits to EPA a plan for offsetting those excess emissions in the next year (or longer if EPA approves). Further, EPA must deduct allowances equal to the excess tonnage from the source's allocation for the next year.

Another EPA responsibility is to provide for allowance auctions. For the post-2000 period, the law sets aside a percentage of available allowances for auction. Anyone may participate in these auctions as a buyer or seller, and those selling allowances may specify a minimum sale price. EPA may delegate or contract the conduct of the auctions to other agencies, such as to the Department of the Treasury, or even to nongovernmental groups or organizations. Two streams of allowances are sold in the auctions. The first stream represents "spot sales" of allowances that must either be used in the year they are sold or banked for use in a later year. The second stream represents "advance sales" of allowances that must either be used in the seventh year after the year they are first offered for sale or be banked for use in a later year. For 2000 and thereafter, Title IV provides that 125,000 allowances be set-aside annually for spot sales, and 125,000 for advance sales.

Administering the Program: The Environmental Protection Agency (EPA)

It is EPA's responsibility to administer the trading, banking, and auctioning of allowances.

Allowance Accounting

EPA has developed an integrated system to track allowances (the Allowance Tracking System— ATS)[25]; to verify and record SO_2 emissions from affected units (the Emission Tracking System— ETS); and to reconcile (true-up) allowances and emissions at the end of the year. The Allowance Tracking System is the official record of allowance transfers and balances used for compliance purposes. Each participant in the system has an ATS account, and each account has an identification number.

Table 1 identifies what the ATS tracks and does not track with respect to allowance activity. As suggested, EPA primarily gathers information to ensure compliance with the emission limitations of Title IV—the ATS is not a trading platform. Participants are not required to record all transfers with EPA until the affected allowances are to be used for compliance. Participants must notify EPA to have any transfers recorded in the ATS. When parties agree on a transaction that they want recorded on the ATS, they provide information on the buyer and seller and the serial numbers of the affected allowances to the ATS which records the transfer.

To facilitate its primary compliance responsibility, EPA assigns each allowance allocated a unique 12-digit serial number that incorporates the first year it can be used for compliance purposes. These allowances may be held in one of two types of ATS accounts. First, there are Unit Accounts where allowances provided under Title IV allocation formulas are deposited and where allowances are removed by EPA for compliance purposes. Second, there are General Accounts that may be created by EPA for anyone wishing to hold, trade, or retire allowances. Participating entities with General Accounts include (1) utilities who keep a pooled reserve of allowances not needed immediately for compliance (i.e., an allowance bank); (2) brokers who need a holding account for allowances in the process of being bought or sold; (3) investors holding allowances for future sale; and (4) environmental and other groups holding allowances they wish to remove from the market (i.e., retire).

Table 1. Information Recorded by EPA's Allowance Tracking System

ATS Records	ATS Does Not Record
Allowances issued	Allowance prices
Allowances held in each account	Option trades
Allowances held in various EPA reserves	Any allowance transaction not officially reported to EPA
Allowances surrendered for compliance purposes	
Allowances transferred between accounts	

Allowance Auctions

A key provision of Title IV to ensure liquidity in the SO_2 markets for new entrants is the EPA allowance auction. As noted above, the EPA is required to auction 250,000 allowances annually in two streams, spot and advance. The auctions began in 1993 and are held annually—usually on the last Monday in March. Sealed bids entailing the number, type, and price, along with payment, are sent to EPA no later than three business days before the auctions.

The auctions sell the allowances according to bid price, starting with the highest bid and continuing down until all allowances are sold or there are no more bids. Unlike allowances offered by private holders for auction, these EPA allowances do not have a minimum price.

For the first 13 years, the auctions were conducted by the Chicago Board of Trade (CBOT) for EPA. CBOT received no compensation for the service, nor was it allowed to charge fees. Beginning in March 2006, CBOT decided to stop administering the auctions; EPA now conducts them directly.

Interface with Electricity Regulation: The Federal Energy Regulatory Commission (FERC) and State Public Utility Commissions (PUCs)

Background

The 1990 Clean Air Act Amendments were enacted during a time of transition in the electric utility industry. There are three components to electric power delivery: generation, transmission, and distribution. Historically, electricity service was defined as a natural monopoly, meaning that the industry had (1) an inherent tendency toward declining long-term costs, (2) high threshold investment, and (3) technological conditions that limited the number of potential entrants. In addition, many regulators considered unified control of generation, transmission, and distribution the most efficient means of providing service. As a result, most people (about 75%) were served by vertically integrated, investor-owned utilities.

The Public Utility Holding Company Act (PUHCA)[26] and the Federal Power Act (FPA) of 1935 (Title I and Title II of the Public Utility Act)[27] established a regime for regulating electric utilities that gave specific and separate powers to the states and the federal government. Essentially, a regulatory bargain was made between the government and utilities. Under this bargain, utilities must provide electricity to all users at reasonable, regulated rates in exchange for an exclusive franchise service territory. State regulatory commissions address intrastate utility activities, including wholesale and retail rate-making. Authorities of these commissions tend to be as broad and varied as the states are diverse. At the least, a state public utility commission will have authority over retail rates, and often over investment and debt. At the other end of the spectrum, the state regulatory body will oversee many facets of utility operation. Despite this diversity, the essential mission of the PUC is the establishment of retail electric prices. This is accomplished through an adversarial hearing process complete with attorneys, briefs, witnesses, etc. The central issues in such cases are the total amount of money the utility will be permitted to collect (revenue requirement) and how the burden of the revenue requirement will be distributed among the various customer classes (rate structure).[28] This is commonly known as "rate of return" (ROR) regulation.

Under the regime set up by the Federal Power Act (FPA), federal economic regulation addresses wholesale transactions and rates for electric power flowing in interstate commerce. Historically, federal regulation followed state regulation and is premised on the need to fill the regulatory vacuum resulting from the constitutional inability of states to regulate interstate commerce. In this bifurcation of regulatory jurisdiction, federal regulation is limited and conceived to supplement state regulation. The Federal Energy Regulatory Commission (FERC) has the principal functions at the federal level for the economic regulation of the electricity utility industry, including financial transactions, wholesale rate regulation, transactions involving transmission of unbundled retail electricity, interconnection and wheeling of wholesale electricity, and ensuring adequate and reliable service. In addition, until passage of the 2005 Energy Policy Act (EPACT05),[29] the Securities and Exchange Commission (SEC) regulated utilities' corporate structure and business ventures under PUHCA to prevent a recurrence of the abusive practices of the 1920s (e.g., cross-subsidization, self-dealing, pyramiding, etc.).

This comprehensive, cost-based approached to regulation began to undergo change in the 1970s and 1 980s as passage of the Public Utility Regulatory Policies Act of 1978 (PURPA)[30] and the Fuel Use Act of 1978 (FUA)[31] helped establish independent electricity generators—electricity producers who sold at wholesale and had no exclusive franchise area. Building on the perceived success of these independent generators under PURPA, the Energy Policy Act of 1992 (EPACT92) created a new category of wholesale electric generators called Exempt Wholesale Generators (EWGs) that are not considered utilities and not regulated under PUHCA.[32] EWGs, also referred to as *merchant generators*, were intended to create a competitive wholesale electric generation sector. EPACT92 effectively initiated deregulated wholesale generation by creating a class of generators that were able to locate beyond a typical service territory with open access to the existing transmission system. EPACT05 continued this process by adding provisions to address system reliability, repeal PUHCA, and modify PURPA.[33]

The current status of these initiatives and resulting state responses is a mixture of states with traditional, comprehensive ROR regulation of electricity and those with a restructured industry with segmented generation, transmission, and distribution components. Over the past 20 years, some States have truncated their ROR regulation to the extent they have chosen to restructure their industry in response to Federal initiatives. In states that have not restructured, the system operates as it has since enactment of the Federal Power Act, with retail consumers paying one price that includes transmission, distribution, and generation. This is referred to as a *bundled transaction*. In states that have restructured, consumers are billed for separate transmission, distribution, and generation charges. This is referred to as *unbundled electricity service*. In those states, retail consumers are allowed to choose their retail generation supplier; however, few states actually have competitive markets for retail choice (exceptions include Texas and Massachusetts). FERC regulates all transmission, including unbundled retail transactions.[34]

FERC Allowance Accounting

With the restructuring of the electric utility industry, FERC generally does not set cost-based rates for electricity generation under its jurisdiction. Rather, FERC conducts a two-pronged horizontal and vertical market power analysis to determine an entity's eligibility for "market-based" wholesale rates.[35] If eligible, the entity may set its wholesale prices according to market demand, not according to production costs.

Because of the market-based nature of FERC wholesale rates, allowances are an accounting issue, not a ratemaking issue for FERC. Electric public utilities and licensees within FERC jurisdiction are required to maintain their books and records in accordance with FERC's Uniform System of Accounts (USofA).[36] The USofA guides the jurisdictional entity in understanding the information it needs to report on various FERC forms. Included in the USofA are instructions on how to account for allowances allocated to the entity under the 1990 Clean Air Act, or acquired by the entity for speculative purposes. Allowances owned for other than speculative purposes are accounted for at cost in either Account 158.1 (Allowance Inventory), or Account 158.2 (Allowances Withheld) as appropriate. Allowances acquired for speculative purposes are accounted for in Account 124 (Other Investments).[37]

By defining allowance value in terms of historic costs, allowances allocated by EPA to entities are valued at zero. FERC does require that the records supporting Account 158.1 and 158.2 be maintained "in sufficient detail so as to provide the number of allowances and the related cost by vintage year." Likewise, the Uniform System of Accounts also provides instruction on accounting for gains and losses from selling allowances.

It should be noted that the Internal Revenue Service (IRS) also values allowances allocated by EPA to an entity on a zero-cost basis.[38]

State Public Utility Commissions

In states with bundled rates, the valuing and disposition of allowances is more than an accounting issue, it is also a ratemaking issue. During and after passage of Title IV, there was substantial debate and studies were done on the role of the PUCs in facilitating (or hindering) allowance trading.[39] In Title IV, the regulatory treatment of allowances is left to the appropriate state and federal regulatory bodies. Title IV contains no mandated requirements regarding the treatment of allowance transactions in state utility rate proceedings. Basically, Congress chose to leave the state commissions free to apply any rate treatment they deem reasonable and appropriate.

The states responded in a diverse manner, some states issuing broad guidelines on treatment of allowance transactions while others decided such events on a case-by-case basis. An analysis of the interaction between PUCs and the allowance system made three general observations about the resulting PUC treatment of allowances: (1) regulations tend to require 100% of both expenses and revenues from allowances to be returned to ratepayers with net gains (losses) incurred used to offset (or increase) fuel costs; (2) a few states have allowed utilities to retain some of the profits as an incentive to sell excess allowances; (3) state regulations tend to be tailored to a state's specific circumstance—"allowance rich" states have regulations encouraging sales, "allowance poor" states have regulations encouraging purchases.[40] The focus of PUC decisions has not been to encourage allowance transactions, but generally to ensure ratepayers and not shareholders receive the benefits of the allowances. In some cases, PUCs have also used their authority to encourage utilities to protect high-sulfur coal production, even if it is not the most cost-effective control strategy.[41]

Allowance Transactions

Internal Transfers

When the 1990 Clean Air Act Amendments were enacted, about 75% of the allowances were allocated to vertically integrated, ROR regulated entities. Today, that percentage has shifted with more allowances allocated to independent generating entities as some utilities have divested themselves of their generating assets. This diversification of ownership is reflected to some degree in the ATS statistics on official transfers and transactions.[42] As indicated by Table 2, in the first two years of trading, transfers between economically unrelated entities were a small percentage of total transfers. More recent data suggest that transfers between unrelated entities account for about 50% of total transfers. However, it is clear that internal transfers remain a major part of the allowance market, even in a restructured industry, and that the total number of official transactions occurring is quite modest.

Internal transfers (i.e., transfers within or between economically related entities) tend to be transacted in accordance with agreements that the utility and/or holding company has filed with the appropriate state PUC, or FERC, or both.[43]

Over the Counter: Cash Market, Futures and Options

Beyond restructuring, other entities are emerging as participants in the allowance markets. This increased diversity of interest in the allowance market is reflected in the recent EPA allowance auctions. For example, as indicated by Table 3, JP Morgan Ventures Energy Corporation bought a substantial position in the auction. This may suggest an increasing importance of intermediaries to the functioning of the allowance market, the development of a more liquid market, and to the maturing of that market.

The basic market for allowance trading is the Over-The-Counter (OTC) market. The most common trading structure involves spot sales with immediate settlement accounting and delivery into EPA's Allowance Tracking System (ATS) with payment by wire transfer in three business days.[44] Daily spot trading volumes for immediate settlement are estimated in the 10,000 to 25,000 ton range.[45] Forward settlement transactions are less common and are fairly short-dated—6 to 18 months out. Vintage swaps also occur in both markets with the difference in value usually paid in additional allowances rather than cash.[46] This preference for allowances reflects regulated entities' desire to keep these transactions non-taxable under current IRS regulations. Cash market transactions are facilitated in some cases through available electronic trading platforms, such as Intercontinental Exchange, Inc. (ICE) and TradeSpark (CantorCO_2e), and by the emergence of a number of allowance brokers. Currently, EPA lists 14 allowance brokers on its website.[47] A similar list is available from the Environmental Markets Association—a trade association.[48] Brokers tend to be registered with the SEC and one or more Self-Regulatory Organizations, such as FINRA; but participation in this market would not in itself make a firm subject to SEC regulation. Four brokers—Cantor Fitzgerald, Evolution, ICAP Energy, and TFS Energy—form the basis of the *Platts* emission price index. *Argus AIR Daily* also produces price indices through daily phone surveys of active brokers.

Two exchanges provide SO_2 future contracts as well as clearing services: New York Mercantile Exchange (NYMEX) and Chicago Climate Futures Exchange (CCFE). The availability of exchanges as a trading platform for allowances or to clear transactions was cheered by traders when established in late 2004 and 2005. As stated by the Environmental Markets Association with respect to NYMEX's decision: "NYMEX does offer information on power, and any time you have them expanding into our market, that's going to create opportunities for people who may be using other products to take a second look at emissions."[49] Both exchanges offer standardized and cleared futures contracts, along with clearing services for off-exchange transactions. As reported by *Platts*, futures volume on both exchanges have expanded greatly over the past year. SO_2 futures trading on the CCFE was nearly 1.9 million allowances in the first half of 2007, compared with about 500,000 during the same time in 2006. For the NYMEX, volumes in the first half of 2007 was 665,000 allowances—a more than three-fold increase over the first half of 2006.[50] Table 4 summarizes the basic features of the trading instruments.

In April, 2007, the CCFE began offering SO2 options.[51] For February 2010, the CCFE offered European-style options[52] on its futures contracts for expiration on the February 2010 through December 2015 futures contracts.[53] As with the futures market, participants are required to settle their delivery obligations via the ATS. In March, 2008, the NYMEX also began offering European-style options on its futures contracts.[54]

Table 2. EPA Official Allowance Transfers and Transactions: 1994-2008

Year	Total Transfers (millions of allowances)	Transfers between economically distinct organizations (millions of allowances)	Percent of Total Transfers	Total Number of Transactions	Transactions Between economically distinct organizations	Percent of Total Transactions
1994	9.2	0.9	9.8%	215	66	30.7%
1995	16.7	1.9	11.4%	613	329	53.7%
1996	8.2	4.4	53.7%	1,074	578	53.8%
1997	15.2	7.9	52.0%	1,429	810	56.7%
1998	13.5	9.5	70.4%	1,584	942	59.5%
1999	18.7	6.2	33.2%	2,832	1,743	61.5%
2000	25.0	12.7	50.1%	4,690	2,889	61.6%
2001	22.5	12.6	56.0%	4,900	2,330	47.6%
2002	21.4	11.6	54.2%	5,755	2,841	49.4%
2003	16.5	8.1	49.1%	4,198	1,544	36.8%
2004	15.3	7.5	49.0%	20,000	n/a	n/a
2005	19.9	10.0	50.3%	5,700	n/a	n/a
2006	22.4	9.5	42.4%	6,400	n/a	n/a
2007	16.9	9.1	53.8%	4,700	n/a	n/a
2008	13.9	5.9	42.4%	3,236	n/a	n/a

Source: U.S. Environmental Protection Agency, 2009.

Table 3. EPA 2009 Auction Results (Winners of More than 20 Allowances)

Spot Market Bid Winners	Quantity	Percent of Total Allowances offered (125,000)
Ohio Valley Electric Corporation	75,000	60.00%
EMMT	48,415	38.73%
Macquarie Cook Power Inc.	1,500	1.20%
Denver City Energy Associates	60	0 .05%
TOTAL SPOT	124,975	99.98%
7 Year Advance Bid Winners	**Quantity**	**Percent of Total Allowances offered (125,000)**
JP Morgan Ventures Energy Corporation	124,984	99.99%
TOTAL ADVANCE	124,984	99.99%

Source: Environmental Protection Agency, 2009

Table 4. SO_2 Futures Contract Specifications

	NYMEX	CCFE
Trading Platform	ClearPort	ICE
Clearing Organization	NYMEX ClearPort Clearing	The Clearing orporation (CCorp)
Self Regulatory Organization	NYMEX and National Futures Association (NFA)	National Futures Association (NFA)
CFTC Regulatory Status	Designated Contract Market	Designated Contract Market
Contract size	100 SO_2 allowances	25 SO_2 allowances
Minimum Price Fluctuation	$25 per contract	$2.50 per contract
Settlement	Physical through EPA's ATS	Physical through EPA's ATS
Symbol	RS	SFI

Source: NYMEX and CCFE.

Regulation of Allowances as an Exempt Commodity: Commodity Futures Trading Commission (CFTC)

Definition

The Commodity Exchange Act provides the basis for federal regulation of "derivative" transactions in contracts based on commodity prices. Pursuant to the act, the Commodity Futures Trading Commission (CFTC) regulates the futures exchanges, such as NYMEX and CME, and certain other derivative transactions that occur off-exchange. The CFTC's authority varies according to the identities of the market participants and the nature of the underlying commodity. In general, the CFTC does not regulate spot (or cash) trades in commodities, or forward contracts that will be settled by delivery of the physical commodity (which are also considered cash sales).[55]

In terms of allowances, the CFTC's jurisdiction is confined to trades that take place on those markets it regulates. It has no jurisdiction over spot trades in allowances, full jurisdiction over futures and options trades on regulated exchanges, and limited jurisdiction over derivatives trades on certain other markets subject to lighter regulation than the exchanges.

Allowances are regulated by the CFTC as exempt commodities under the Commodity Futures Modernization Act of 2000.[56] The Commodity Exchange Act defines an exempt commodity as any commodity other than an excluded commodity (e.g., financial indices, etc.) or an agricultural commodity. Examples include energy commodities and metals. Emission allowances are related to energy production. This designation has been supported by other federal entities. In a 2005 Interpretive Letter approving physically settled emission derivatives transactions, the Office of the Comptroller of the Currency, Administrator of National Banks, states that physical settlement of emission allowances do not pose the same risk as other physical commodities:

> The proposed emissions derivatives transactions [e.g., futures, forwards, options, swaps, caps, and floors] will be linked to three emission allowance markets: the U.S. SO_2 (Sulfur Dioxide) and NOx (Nitrogen Oxide) markets and the European Union's CO_2 (carbon dioxide) market. These emissions markets are volatile and price fluctuates considerably. Market participants manage price risk through the use of derivative structures, such as forwards, futures, options, caps and floors. These derivatives are generally physically settled, because the current emissions market is primarily physical in nature....
>
> The OCC has previously concluded in a variety of contexts that national banks may engage in customer-driven commodity transactions and hedges

> that are physically settled, cash- settled and settled by transitory title transfer.... Similarly, the OCC permitted a national bank to make and take physical delivery of commodities in connection with transactions to hedge commodity price risk in commodity linked transactions....
>
> In these decisions, the approved activities were subject to a number of conditions due to risks associated with physical transactions in certain commodities. Those risks included storage (e.g., storage tanks, pipelines), transportation (e.g., tankers, barges, pipelines), environmental (e.g., pollution, fumigation, leakage, contamination) and insurance (e.g., damage to persons and property, contract breach, spillage). Physical settlement of emissions derivatives and hedging with physicals would not pose those risks, however. Emission allowances are not tangible physical commodities, such as electricity or natural gas. Rather, they are intangible rights or authorizations. They can be bought and sold like other commodities, but they exist only as a book entry in an emissions account.[57] [footnotes omitted]

The Federal Reserve also considers emission allowances as commodities for purposes of trading.[58]

Regulation of Trading Venues

The CFTC identifies four venues for trading exempt commodities under the Commodity Exchange Act: (1) Designated Contract Markets (DCM), (2) Commercial Derivatives Transaction Execution Facilities [none currently in operation], (3) Exempt Commercial Markets (ECM), and (4) Over-the-Counter (OTC)—not on a trading facility.[59] As suggested by the discussion above, allowances are traded on three of these venues. Futures contracts and clearing services are provided by NYMEX and CCFE—both DCMs—with options also available on the CCFE. ICE and TradeSpark—both ECMs—are used by brokers and principals for allowance transactions. Finally, principal-to-principal transactions and broker-assisted transactions are occurring OTC without the use of a trading facility. Table 5 summarizes these venues and their regulation under the Commodity Exchange Act.

For the three trading venues set out in Table 5, the degree of regulation varies, most significantly according to the identities of the participants. Small public investors are allowed to trade only on regulated exchanges (DCMs); these are subject to extensive self-regulation and CFTC oversight. Electronic trading facilities, where small traders are not present, are subject to much less regulation, because traders are assumed to be capable of protecting themselves from fraud. However, if an electronic trading facility plays a significant price discovery role (that is, if the prices it generates are used as reference points by the cash market or other derivatives markets), the CFTC may require disclosure of certain information about trading volumes, prices, etc. Where trades are purely bilateral, negotiated, and executed between principals, the transaction is said to occur in the OTC market, which is entirely exempt from CFTC regulation, with the exception of certain provisions dealing with fraud manipulation.

Table 5. Summary of Trading Venues for Exempt Commodities under the Commodity Exchange Act (CEA)

	Designated Contract Markets (CEA Sec. 5)	**Exempt Commercial Markets (CEA Sec. 2(h)(3)-(5))**	**OTC—Not on a Trading Facility (CEA Sec. 2(h)(1)-(2))**
Commodities Permitted	No limitations	Exempt commodities (e.g., energy metals, chemicals, emission allowances, etc.)	Exempt commodities (e.g., energy metals, chemicals, emission allowances, etc.)
Method of Trading	Trading can take place on an electronic trading facility or by open outcry	Electronic multi-lateral trading (i.e., many-to-many platforms)	Non-multi-lateral trading (e.g., dealer markets; individually-negotiated, bilateral transactions)
Notice Requirement	Must apply to and receive prior approval from CFTC; must satisfy various non-prescriptive designation criteria and core principles	Yes; simple notice containing contact information and description of operations	None; exemption is self-executing
Participants	No limitations	Eligible Commercial Entities only—subset of Eligible Contract Participants; excludes individuals but includes funds	Eligible Contract Participants (i.e., institutions, finds, and wealthy, sophisticated individuals)

Table 5. (Continued)

	Designated Contract Markets (CEA Sec. 5)	Exempt Commercial Markets (CEA Sec. 2(h)(3)-(5))	OTC—Not on a Trading Facility (CEA Sec. 2(h)(1)-(2))
Intermediation	Permitted	None; principal-to-principal trading only	Limited; only if done through another Eligible Contract Participant
Types of Transactions	Futures and options	Derivatives, including swaps, futures and options (Note: ECMs often also trade products outside CFTC jurisdiction, including spot and forward contracts)	Derivatives, including swaps, futures, and options
Standardized Products?	Yes	Yes, terms set by the entity	Usually yes when executed on a dealer market. Usually no, when executed bilaterally
Cleared?	Transactions must be cleared through a Derivatives Clearing Organization (DCO) approved by the CFTC	Clearing not mandatory; if offered, it must be through an SEC-registered clearing agency or a DCO (many ICE transactions are cleared at LCH; other ECMs offer clearing at NYMEX Clearport or The Clearing Corp.)	Can be if a standardized contract; many traders choose to clear trades at NYMEX or LCH
Transaction Prohibitions	Subject to all provisions of the CEA	Only anti-manipulation and anti-fraud	Only anti-manipulation and anti-fraud (but

Table 5. (Continued)

	Designated Contract Markets (CEA Sec. 5)	Exempt Commercial Markets (CEA Sec. 2(h)(3)-(5))	OTC—Not on a Trading Facility (CEA Sec. 2(h)(1)-(2))
			anti-fraud rules do not apply to transactions between Eligible Commercial Entities)
Self-regulatory responsibility	Yes, significant self-regulatory responsibilities; must comply on a ongoing basis with 8 designation criteria and 18 core princeples. Must have compliance and surveillance programs	Minimal and they include nothing that goes to the integrity of trading. Responsibilitiesinclude a reporting requirement for contracts over a minimum volume threshold; ensuring compliance with exemption conditions; and dissemination of contract activity information for "price discovery" contracts	None
Responsibility to CFTC	Comply with designation criteria and core principles	Provide notice of operation and weekly transaction data for high-volume contracts; report manipulations and fraud complaints; maintain and provide access to records of activity	None

Table 5. (Continued)

	Designated Contract Markets (CEA Sec. 5)	Exempt Commercial Markets (CEA Sec. 2(h)(3)-(5))	OTC—Not on a Trading Facility (CEA Sec. 2(h)(1)-(2))
CFTC Oversight Authority	Unlimited, inclu-ding continuous and ongoing market surveillance and trade practice programs, ability to intervene in markets (e.g., force reduction/liquidati ons of position, alter/supplement DCM rules). CFTC receives large trader reports transaction data and assesses DCMs' compliance programs via rule enforcement reviews	Limited (special calls); Sec 8a(9) emergency auth-ority does not apply	None

Source: *Venues for the Trading of Exempt Commodities under the Commodity Exchange Act (CEA)*, available on the CFTC website at http://www.cftc.gov/*stellent/groups/public/@newsroom/documents/file/exemptcommoditiesvenues_09 1 207.pdf.*

Although allowances are regulated like any other commodity by the CFTC, it should be noted that it is not a deep liquid cash market. As noted by emissions broker Evolution Markets LLC, the affected source base for SO_2 allowances is about 500 companies. The broker also estimated in 2005 that about 20 companies represented the bulk of trading activities.[60] In recommending CFTC approval of the CCFE as a DCM, the Staff memorandum noted the following:

> In futures markets generally, the existence of a liquid market for a particular contract and the ability of an FCM to liquidate positions therein which it may inherit from a defaulting customer are important to the financial

integrity of such an FCM and, in turn, its ability to fulfill its obligations to other customers and to the clearing system. The EPA will facilitate the delivery process of these contracts in a manner that makes cash positions known and compensates for any current lack of a developed deep liquid cash market for the contracts as compared to other futures contracts. Collectively CCorp, NFA, and EPA will carry out financial surveillance, monitor situations, and provide information the effect of which should counterbalance any disparate effects on financial integrity, which might be imposed by the initial lack of trading history and prices.[61]

Derivatives Reform Proposals

In the wake of the financial crisis that erupted in 2007, the current regulatory regime for OTC derivatives was singled out as a weak link in financial system oversight. Legislative proposals passed by the House (H.R. 4173) and before the Senate (the Dodd committee print on regulatory reform and legislation being developed in the Agriculture Committee) would require more reporting, capital, and clearing of many OTC contracts. There are significant differences among the proposals, but all appear to agree on certain basic reforms:

- all swaps will be reported to a swap repository, so that regulators will know who is trading what contracts and in what volumes;
- many contracts will be moved onto clearing houses where traders will have to post margin or collateral to cover potential and actual losses (all proposals contain exemptions for "end-users," or firms that use derivatives to hedge commercial risk. Many users of carbon derivatives would presumably fall into the exempt end-user category);
- certain contracts will have to be traded on exchanges or some form of regulated swap execution facility;
- the distinction between exempt and other commodities will be repealed—all nonagricultural derivatives will be subject to comparable regulatory requirements, regardless of the underlying interest; and
- regulators will set limits on the size of speculative positions in derivatives, to prevent price manipulation.

Under these derivatives reform proposals, CFTC would remain the regulator of carbon derivatives, whether exchange-traded or OTC. Carbon contracts would not be subject to any special regulatory treatment.

Lessons and Observations from Title IV Program

Despite the tendency to view the Title IV program as a model for a future greenhouse gas reduction scheme, there are several important differences. For example, the Title IV program involves up to 3,000 new and existing electric generating facilities that contribute two-thirds of the country's SO_2 and one-third of its nitrogen oxide (NOx) emissions (the two primary precursors of acid rain). This concentration of sources makes the logistics of allowance trading administratively manageable and enforceable with continuous emissions monitors (CEMs) providing real time data. However, greenhouse gas emissions are not so concentrated. In 2005, the electric power industry accounted for about 33% of the country's GHG emissions, while the transportation section accounted for about 28%, industrial use about 19%, agriculture about 8%, commercial use about 6%, and residential use about 5%.[62] Thus, small dispersed sources in transportation, residential/commercial and agricultural sectors, along with industry, are far more important in controlling GHG emissions than they are in controlling SO_2 emissions. This diversity multiplies as the global nature of the climate change issue is considered, along with the multiple GHGs involved.[63] Thus, a carbon market is like to involve far greater numbers of affected parties from diverse industries than the current Title IV program.

It will also involve far greater numbers of tradeable allowances than the current Title IV program. Under the current program, about 9 million allowances are allocated to participating entities annually. In contrast, a domestic greenhouse gas program that capped emissions in the electric power, transportation, and industry sectors at their 1990 levels at some point in the future would be allocating about 4.85 billion allowances annually. This is a two and a half orders-of-magnitude increase over the Title IV program and double the Phase 2 allocations under the ETS. Trading activities under Title IV has been increasing since 2005; however, the volumes don't approach those anticipated if a greenhouse gas cap-and-trade program were instituted.

Finally, the economic value of a future carbon market is likely to be substantially greater than the Title IV program. In 2009, the auction price for a Title IV allowance was $62. Thus, the annual allocation of SO_2 allowances has a market value of about $0.5 billion. Using estimates of $15 to $25 an allowance, the annual allocation of 4.85 billion allowances posited above for a greenhouse gas program would have a market value of $72.8 billion to $121.3 billion.[64] Unlike the Title IV market, a carbon market may be quite liquid, particularly as the market expands globally.

Despite these differences in scope and magnitude, there are trends in Title IV trading that are likely to continue in a carbon market.

First, there is a trend toward more diverse, non-traditional participants in the Title IV market. Like the Title IV market, the economic importance of a carbon market will likely draw in entities not directly affected by the reduction requirements, such as financial institutions. The motivations of these entities may be equally diverse, including facilitating projects involving the need for allowances, portfolio balancing, and profits earned through intermediary fees or proprietary trading.

Second, there is trend in the Title IV market toward using financial instruments to manage allowance price risk. This trend is partly the result of the regulatory uncertainty introduced in the allowance market by difficulty EPA is having with the Clean Air Interstate Rule (CAIR). Given the greater economic stakes involved in a carbon market, this trend toward more sophisticated financial instruments is likely to emerge early as a hedge against price uncertainty. The emergence of entities well-versed in the use of these instruments may reinforce the trend and make options, collars, strangles, and other structures as common in the allowance market as they are in other commodity markets. With a more liquid and dynamic market, a carbon market may look more like other energy markets, such as natural gas and oil, than the somewhat sedate SO_2 allowance market.

IMPLICATIONS FOR A FUTURE U.S. CARBON MARKET: REGULATORY ISSUES

If the United States adopts a cap-and-trade system based on CO_2 allowances, the resulting trading market would be large. As noted above, between four and five billion allowances might be issued each year, with a market value of around $70 to $120 billion. Judging by the interest already expressed by major Wall Street firms, the value of secondary market trading might be several times that figure.

Since thousands of businesses would be affected by a mandatory emissions trading system, there is a strong public interest in ensuring that (1) the market functions smoothly and efficiently, generating prices that accurately reflect supply and demand for emissions permits, (2) the market is free from fraud and manipulation, which could potentially arise from a number of sources, including market power in the hands of a few firms and the abuse of nonpublic, price-sensitive information, and (3) market participants' and regulators' need for transparency is balanced against legitimate business concerns about the release of confidential, proprietary information.

An Efficient Trading and Pricing Mechanism

A generation ago, only a few large, established exchanges would have had the capacity to handle the trading volumes anticipated for a U.S. CO_2 market. That is no longer the case: thanks to cheap computing power and telecommunications, small firms with a few dozen employees can handle much of the volume of the major stock exchanges, which employ thousands. Thus, we can expect trading mechanisms to emerge quickly once U.S. carbon trading is authorized, with no need for government assistance.

A number of organizations and groups have been preparing for the advent of emissions trading. The Chicago Climate Exchange (CCX) already operates a voluntary greenhouse gas emissions market, featuring both spot and futures trading.[65] Cantor Fitzgerald, a securities firm that operates a secondary trading platform for U.S. Treasury securities, has formed a subsidiary, CantorCO2e, which offers "an integrated set of services—transaction, financing, technology and consultancy— to bring environmental commodities to market and to assist clients across the world in managing the financial aspects of energy and environmental choices."[66] In December 2007, the New York Mercantile Exchange (Nymex), the leading U.S. energy futures market, announced the formation of the Green Exchange, in a joint venture with several major investment banks. (Nymex has since merged with the Chicago Mercantile Exchange, or CME Group.) The Green Exchange offers environmental futures, options, and swaps, and expects to register with the Commodity Futures Trading Commission (CFTC) as a futures exchange.[67] Intercontinenta-lExchange (ICE) already owns and operates a CFTC-regulated futures exchange (the former New York Board of Trade), which specializes in sugar, cotton, and other agricultural commodities but could easily offer emissions-related contracts as well. Any of these entities would have little difficulty in hosting large-scale trading of emissions allowances.

Since start-up costs are relatively low, it is likely that a number of competing trading venues would emerge to handle U.S. emissions trading if the United States followed the EU's example and did not mandate a particular trading mechanism or structure. Competition would favor markets with low trading costs, easy access, and fast and reliable execution systems. The usual pattern in trading markets is for volume to gravitate to a single market, to enhance liquidity.[68]

This suggests that the less successful competitors might be relegated to niche markets, that a number of trading mechanisms could be linked electronically to form a single market, or that mergers and acquisitions would reduce the number of trading venues.

It is of little public interest which firm or firms emerges as the market leader. What is important, however, is that there be an authoritative source of price information, since the price of emissions will guide firms considering investment in pollution abatement. Price discovery mechanisms are most efficient in liquid markets, where many traders bring information to the price-setting process by their buying and selling decisions.

In general, the functioning of the price mechanism has not been a concern of financial regulators.[69] The Securities and Exchange Commission (SEC) and the CFTC both rely on self- regulation by the exchanges in this area, reflecting the fact that the exchanges predate the federal regulators by decades or centuries. Both agencies have broad authority to intervene if they determine that prices do not accurately reflect the underlying forces of supply and demand.

If patterned after the Title IV program and the EPA maintained the registry of allowance ownership, that agency might play a role in price dissemination in a future carbon market. Reports to the EPA that a trade has occurred could be required to include price information, which could be published electronically. In practice, however, the EPA might not be the best source of price data. Since the EPA would presumably receive only spot market trade data, and since most proposals follow the Title IV procedure that specifies that each transaction report to the EPA must include a written certification of the transfer, signed by a responsible official of each party, the question arises whether prices generated by the EPA registry would be outdated by the time they were published. In practice, many commodity spot markets look to the futures exchanges for current prices; new prices are generated there second-by-second, recorded, and transmitted almost instantaneously. If substantial numbers of spot transactions took place on exchanges, as they do in Europe, current information on spot prices might also be available.

Regulators might have a role to play in ensuring that the exchanges did not charge excessive fees for access to real-time price data. Both the SEC and CFTC have grappled with this issue— customers tend to view the exchanges as public utilities, and prices as common goods, but the exchanges naturally look upon price data as their private property, to be sold for what the market will bear.

The availability of price information also depends on the regulatory status of the source market. Exchanges regulated by the SEC or CFTC are required to disclose price and volume data. (The proprietary issue mentioned above refers only to intra-day or real-time prices.) These rules do not apply with the same force to over-the-counter (OTC) markets, which are a significant presence in financial and energy derivatives markets, but which are largely exempt from CFTC regulation.[70] Under current law, the CFTC has very limited authority to require OTC markets to disclose trading data.

Derivatives reform proposals in the 111th Congress would mandate more price transparency in OTC markets, where contract terms are not now made public.

Fraud and Manipulation

Regulation of secondary emissions trading would have to consider two kinds of fraud and manipulation: fraud by traders or intermediaries against other investors, and sustained price manipulation, which is harmful not only to market participants, but potentially to consumers and the economy.[71]

Investor Fraud

Both CFTC and SEC have extensive experience with numerous programs designed to prevent and punish fraud. Much anti-fraud regulation takes place in a self-regulatory framework: regulated exchanges are required to establish and enforce rules to promote fair trading. To choose two examples from a very long list of investor protection rules, securities brokers are bound by a duty of best execution—they are required to obtain the best terms reasonably available to fill a customer's order—while futures commission merchants are prohibited from trading for their own accounts when they have an unfilled customer order in hand. Many similar protections would apply to trades in emissions allowances or derivatives executed on a regulated exchange, whether futures or securities. Both types of markets require professional traders to maintain accurate records of all transactions. There is no strong reason to think that CFTC or SEC regulation would be superior in protecting small traders: either should be adequate.

In addition to federal statutes and regulations and the rules of self-regulating securities and futures exchanges, state laws provide protections against crooked dealings. Brokers, investment advisers, and other intermediaries are regulated at the state level in their transactions with customers, to prevent them from taking advantage of public investors with less knowledge of current market conditions.

The degree to which fraud on the unwary could be a problem depends on how many and what kinds of traders are attracted to the market. In the SO_2 market, it is not a significant problem. SO_2 emissions are primarily generated by relatively few power utilities, each of which has roughly the same information as the others. All see the same weather forecasts and energy price data. In such a market, it is difficult for one party to defraud another.

The CO_2 market would be different: there would be thousands of firms in the market, and potentially millions if the program were to cover transportation emissions. Assuming that significant information asymmetries exist between small firms in the market and large ones, the latter are likely to be in a position to take advantage of the former. Either SEC or CFTC regulation could be appropriate, depending on whether emissions allowances and derivatives came to be traded on securities or futures exchanges.

In OTC derivatives markets, where only sophisticated investors and institutions are allowed to trade,[72] participants are assumed to have the incentive and capacity to protect themselves from fraud.[73] This is the general rationale for exempting certain markets from regulation: if public customers are not present, there is thought to be no public interest in providing investor protection at the taxpayers' expense.

Spot market trades, which would not come under regulation under current securities or commodities law, are another area where abuses could arise from information asymmetries between large, sophisticated traders and smaller firms that rarely use the market. The potential for abuse, however, would be greatly reduced if current price data were easily available to all market participants.

Inside Information

The European experience suggests that there may be opportunities to trade on inside, non-public information. Even sophisticated market participants are vulnerable to this type of fraud. The concepts of insider trading, however, are not the same in securities and futures markets.

Under securities law, insider trading involves the use of nonpublic information about a single firm. Corporate insiders in possession of nonpublic information that is material—that is, that would affect a reasonable investor's decision to buy or sell—are prohibited from buying or selling the company's shares until the information is disclosed to the public. All trades by certain executives, officers, and directors in their own company's shares must be disclosed within two business days. The definition of who is an "insider" has been expanded by legislation and court decisions in recent decades, so that under certain circumstances investment bankers, journalists, and various fiduciaries may be encompassed.[74]

In futures markets, there is no equivalent to inside information about a single corporation's prospects. Contracts are based on homogenous commodities that thousands of people produce, trade, and consume. Certainly some traders—large producers or industrial users of commodities—have information that others do not have, but the price discovery process depends on that information being incorporated into the price.[75]

The Commodity Exchange Act and CFTC regulations bar exchange employees (or governing board or committee members) from trading on material nonpublic information obtained through their positions, or from "tipping" others to trade on such information.[76] This prohibition does not extend to everyone who trades in the market.

Because of the variance between futures and securities concepts of insider trading, if, hypothetically, the CEO of ExxonMobil were to buy stock knowing that the discovery of a major new oil field would be announced the next day, the transaction would be clearly illegal under the securities laws. If, however, in the same circumstances he sold crude oil futures contracts, expecting that the announcement would lower the price of oil, it would not necessarily be unlawful.

Thus, it is not certain that either the CFTC or the SEC approach to insider trading would be adequate to protect the market from the kind of abuse suspected in Europe before the allowance price crashed in 2006 (where traders may have used pre-publication data regarding emissions levels and allowance allocations). What may be required is a hybrid of securities and commodities law, imposing disclosure requirements on certain firms in possession of material nonpublic information.

Market Manipulation

Investor protection is not the only goal of anti-fraud regulation. Even though sophisticated investors in unregulated, private markets have the incentive and means to protect themselves against fraud by those with superior information, these markets may be just as vulnerable to price manipulation as the public markets. In recent years, there has been widespread concern in Congress and elsewhere that excessive speculation in energy derivatives may have caused the prices of oil and natural gas to be higher than the fundamentals of supply and demand would justify.[77]

Allowance price manipulation would resemble the kind of manipulation that the CFTC is equipped and accustomed to prevent, detect, or punish. Allowance prices would be subject to corners and squeezes to the same extent as any commodity contract, assuming that allowances were issued in advance, as they are in Europe, and that the supply could not be expanded before the beginning of the next allocation cycle. To corner the market, a manipulator would amass a large inventory of allowances while simultaneously taking futures or forward positions that required others to make delivery to it. When a squeeze is successful, traders with delivery obligations have no choice but to buy from the manipulator at prices it can dictate, and then sell those same allowances back to the manipulator at the lower prices specified in the futures and forward contracts.

To prevent manipulations of this type, which can cause prices of the underlying commodity (emissions, in this case) to rise far above fundamental levels for extended periods of time, the CFTC has a number of surveillance programs that apply to both the spot and futures markets. First, the CFTC maintains a large trader reporting system: anyone controlling more than a specified number of contracts must report the position daily. This information is not made public, but it allows the CFTC to observe the accumulation of large positions that could serve as the basis for manipulation. CFTC is able to aggregate positions held by a single trader with various brokerage firms. Second, the CFTC monitors the deliverable supply of commodities, particularly as the expiration date of the futures contract draws near. If unusual shortages in deliverable supply emerge, the CFTC can take certain remedial steps.[78] However, the CFTC does not have the capacity (nor a clear statutory mandate) for comprehensive monitoring of spot trading in all the commodities upon which futures contracts are based.

The futures exchanges and clearing houses have strong incentives to prevent manipulation—since futures trading is a "zero-sum" game,[79] victims of manipulated prices will almost certainly include many exchange members. Clearing houses, which guarantee payment on all contracts, face the risk that a squeeze or corner may cause many traders to default on their obligations.

To deter manipulation, the exchanges impose position size limits on certain contracts, but these apply only to speculators. Hedgers, those who produce or deal in the underlying commodity, are generally exempt. In the allowance market, as noted above, a concern is that a single large emitter, or a group acting in concert, could have enough market power to influence prices. If concentration in the allowance market is significantly greater than in other commodities, a system of position limits that applies to hedgers might be useful.

SEC anti-manipulation efforts, on the other hand, are generally not aimed at manipulations of an entire market, but at the schemes to distort the price of a single stock, or group of stocks. Many such manipulations deal with misuse of inside information or spreading false information. There is no general equivalent to the CFTC's large trader reporting system, although buyers of more than 5% of a public company's shares must disclose their ownership and state whether the investment is passive or whether control of the company is sought.

The Energy Policy Act of 2005 (P.L. 109-58) significantly expanded the authority of the Federal Energy Regulatory Commission (FERC), giving the agency a new role in regulating energy markets. The act expanded the jurisdictional reach of FERC, authorizing it to address any manipulative device and any entity participating in or affecting FERC's jurisdictional markets (primarily bulk electricity and interstate natural gas). FERC also received new civil authority to impose penalties of up to one million dollars per violation per day. FERC oversees power markets that involve extensive administrative adjustments to regional markets through market monitors, but not markets that would be characterized as exchanges.

Even before the 2005 Act, beginning in 2002, FERC had built a new analytic capability to examine markets and look for anomalies in response to the Enron and California electricity crises. FERC established a Market Oversight unit, which grew to approximately 50 staff. To comply with its new regulatory mission under the 2005 Act, resources were shifted to emphasize enforcement and audit activities. As a result, the market oversight unit lost staff and has continued to shrink, to perhaps a dozen full time professionals at this time.

Though it is relatively inexperienced as a regulator of secondary markets, FERC might play a key role in coordination with another agency (or agencies). Since the prices of natural gas and wholesale electricity would be affected by emissions costs (and vice versa), the possibility of intermarket price manipulation exists. Without the information available to FERC, it will be difficult to obtain a comprehensive overview of supply and demand in the emissions market.

If an OTC market for allowance derivatives were to develop in the absence of comprehensive derivatives reform, no federal agency would have much authority over it, or much information about trading volumes, prices, and market conditions. The extent to which existing OTC markets in financial and energy markets facilitate manipulation is controversial.[80] It is clear that the exchange and OTC markets are economically linked and that swaps and futures are interchangeable from the trader's point of view, and that therefore prices in one market affect the other. Some argue that OTC manipulation is unlikely because all participants are sophisticated and because manipulation of exchange market prices, which are visible and often used in the spot market, is likely to be more profitable. Others argue that the less-transparent OTC market is where a would-be manipulator might choose to accumulate market power, out of the regulator's sight. In any case, the issue is now squarely before the Congress, and emissions trading is unlikely to present any strategies for manipulation not already found (or suspected) in the energy market.

Transparency versus Confidentiality

In order for the cap-and-trade system to work, firms must have clear price signals to guide their investment decisions. Reliable and transparent prices also foster liquid secondary markets, which in turn can make pricing more efficient.

To public investors and customers, transparency means primarily the availability of timely and affordable price information. Without such information, they are prey to better-informed market insiders. Both SEC and CFTC have well-established standards in this area.

The more difficult issue is how much information the public needs about how prices are set. Confidence in the market depends on investors believing that prices are determined fairly, in response to real economic factors, and not manipulated. Providing that assurance is a major function of the regulators.

But there are limits to transparency; beyond a certain point, it becomes very expensive and may harm the market. The discussion above about the market impact of large trades is one example of this. Large traders prefer to trade anonymously, in order that other traders not jump in and sell when they are selling, or buy when they are buying, raising their transactions costs. In the stock market, a number of electronic trading venues have thrived in competition with the major exchanges by offering anonymity to large institutional traders. The potential danger here, however, is that a fragmented, or two-tier, market may evolve, where market professionals and large traders deal with each other at prices better than those they offer to the public.[81]

Regulators also want information that traders would prefer to keep private. Large positions in futures, as noted above, are reported to the CFTC but not made public. There have been no instances to date of trading abuses involving mishandling of large trader data.

In the emissions market, government agencies would have access to large amounts of price- sensitive data. The EPA, first of all, would have full information about the total number of allowances to be issued, and how they were to be allocated. It would also have data on actual measured emissions, which could send important price signals under certain market conditions.

Safeguarding this kind of information, however, need not present any special regulatory problems. Many federal agencies already generate statistics and forecasts that affect market prices—macroeconomic and agricultural data are examples—and existing procedures and laws appear to work well to prevent pre-publication leaks.

Concluding Remarks

Both the European Union's Emissions Trading Scheme and the U.S. Title IV sulfur dioxide program have insights into regulatory issues that may face any future U.S. carbon market. A review of the initial operations of the EU-ETS indicates some potential pitfalls facing a future U.S. market. The 2006 price crash, by weakening the financial incentive to reduce pollution, undermined the basic rationale for the ETS. Although the problem did not originate in the secondary market, it did raise questions about the adequacy of market regulation. In particular, many observers suspect that information about the size of allocations leaked before official publication, and that certain traders profited from this knowledge. Traders with nonpublic information on certified emissions data—which gives an indication of future demand for allowances—may also have profited.

A review of Title IV's much longer trading history reveals at least two trends that are likely to continue in a carbon market and to challenge regulators. First, there is a trend toward more diverse, non-traditional participants in the Title IV market. Like the Title IV market, the economic importance of a carbon market will likely draw in entities not directly affected by the reduction requirements, such as financial institutions. These entities' motivations may be equally diverse, including facilitating projects involving the need for allowances, portfolio balancing, intermediary fees, and trading profits.

Second, as noted, there is a trend in the Title IV market toward using financial instruments to manage allowance price risk. Given the greater economic stakes involved in a carbon market, this trend toward more sophisticated financial instruments is likely to emerge early as a hedge against price uncertainty. The emergence of entities well-versed in the use of these instruments may reinforce the trend and make options, collars, strangles, and other structures as common in the allowance market as they are in other commodity markets. With a more liquid and dynamic market, a carbon market may look more like other energy markets, such as natural gas and oil, than the somewhat sedate SO_2 allowance market.

Regulation of secondary emissions trading would have to consider two kinds of fraud and manipulation: fraud by traders or intermediaries against other investors, and sustained price manipulation, which is harmful not only to market participants, but potentially to consumers and the economy. Four federal agencies could have roles in the regulation of a secondary market in emissions allowances. Each has attributes that may contribute to effective regulation.

The CFTC is the agency that currently oversees the Title IV program and whose current mission most closely resembles what a regulator of the prospective carbon emissions market would do. Secondary CO_2 trading in the EU does not appear to differ significantly from other commodities markets. CFTC is engaged in intermediary regulation, has experience in market surveillance to prevent or detect fraud and manipulation, and supervises a market system—based on the self- regulating exchange and clearing house—that has withstood severe financial shocks.

The major failing of the CFTC, according to some observers, is that it lacks the resources and the statutory mandate to do its job.[82] Futures trading has grown explosively in recent years; CFTC employment and budget have not kept pace. The OTC derivatives markets represent another major challenge. Many believe that the CFTC's limited jurisdiction over the OTC markets constitutes a regulatory gap, and that excessive, unregulated speculation in energy contracts may be partly to blame for high and volatile energy prices.

The SEC is much larger than the CFTC, but it has fallen short in some cases, such as the detection and prevention of fraudulent investment advisers. While the CO_2 market will resemble commodities markets more closely than securities, the SEC has a number of regulatory tools that might be appropriately applied to the emissions market. These include regulation of insider trading and disclosure of material formation by firms in the market. It is likely that SEC-regulated investment banks will play a significant role in the development of a U.S. carbon trading market, as they increasingly are doing in the Title IV market. It is also possible that CO_2 derivatives contracts or indexes could be listed and traded on securities exchanges.

If patterned after the Title IV program, EPA would be responsible for the primary market in allowances: the original allocation and/or auction of permits to emitting firms. It would maintain the registry of ownership of allowances. EPA lacks experience comparable to that of the CFTC and SEC in regulating trading markets, but the information it gathers in the primary market could be critical to oversight of the secondary market.

FERC was granted oversight authority over bulk electricity and interstate natural gas markets in 2005. Its experience with market surveillance and enforcement is thus limited in comparison to the SEC and CFTC, and it does not play an active role in overseeing the Title IV market. FERC may have a role to play, however, based on its oversight of commodity markets with price linkages to CO_2 emissions, but at this time it may be less well-equipped than the CFTC or SEC to be the primary regulator of secondary trading.

If development of secondary trading is left to market forces, as it was in Europe, it is possible that no single regulator would have clear jurisdiction, as is the case in the Title IV program. CO_2- related contracts, as well as the allowances themselves, could be traded simultaneously on futures and securities exchanges, spot markets where EPA has some jurisdiction, and OTC markets that are essentially unregulated.

This kind of regulatory fragmentation has not always worked well in financial markets. After the stock market crash of 1987 revealed differences of opinion among the CFTC, the SEC, and the Federal Reserve, President Reagan created the President's Working Group on Financial Markets,[83] which remains active, conducting studies and making recommendations on intermarket issues, as well as providing a forum for regulatory coordination. A similar umbrella group might help to prevent regulatory gaps or conflicts in the emissions market.

Appendix. Regulation of EU Emissions Exchanges

European Climate Exchange (ECX)

ECX is recognized as the leading secondary market for emissions trading, but it is not itself an organized financial exchange. ECX contracts are traded on the platform of another market, ICE Futures Europe, a Recognised Investment Exchange in the U.K., supervised by the Financial Services Authority (FSA) under the terms of the Financial Services and Markets Act of 2000.

"Recognition" as an investment exchange is contingent upon meeting FSA standards regarding financial resources, controls over systems and conflicts of interest, investor protections, fair access to trading facilities, trade recording, disclosure, custody of customer funds, disciplinary programs, and prevention of fraud and financial crime.

The parent company, IntercontinentalExchange, Inc., is an American firm headquartered in Atlanta. ICE Futures Europe was previously the International Petroleum Exchange, the leading European energy derivatives market. In addition to ICE Futures, ICE operates an OTC electronic platform, which is registered as an exempt commercial market under the U.S. Commodity Exchange Act and the regulations of the CFTC. The CFTC generally oversees, but does not substantively regulate, the trading of OTC derivative contracts on the ICE platform. All ICE participants must qualify as eligible commercial entities, as defined by the Commodity Exchange Act, and each participant must trade for its own account, as a principal. The U.K.'s FSA does not supervise the OTC emissions market.

Under FSA rules, derivatives trading is a "regulated activity," which can only be carried out on a recognized exchange. Traders also need FSA authorization to participate in derivatives markets if they engage in a "specified activity," including (1) dealing in investments as principal; (2) dealing in investments as agent; (3) arranging deals in investments; and (4) advising on investments. There are exemptions from this authorization requirement for hedgers, or those who use derivatives to mitigate risks in their normal course of (unregulated) business.

ICE operates its sales and marketing activities in the U.K. through ICE Markets which is authorized and regulated by the FSA as an arranger of deals in investments and agency broker.

ECX, through ICE Futures, offers futures contracts and options based on EUAs. One contract represents 1,000 tonnes of CO_2 EU Allowances, or 1,000 EUAs. ECX is not a spot market.

BlueNext

BlueNext was formed in December 2007, when NYSE Euronext (the holding company that owns the New York Stock Exchange) and Caisse des Depot (the French government's investment company, or sovereign wealth fund) purchased the weather and carbon trading operations of Powernext, an exchange that trades spot and futures contracts in electricity and natural gas. BlueNext is currently the leading spot market for EUAs, and began trading EUA futures contracts and contracts based on CERs during 2008.

Like Powernext, BlueNext is classified as an investment company which manages a multilateral trading facility, and is registered with the Comité des Etablissements de Crédit et des Entreprisesd'Investissement under the oversight the Autorité des Marchés Financiers (AMF), or Financial Market Authority. The Commission de Régulation de l'Energie (CRE) and the DIDEME (French Ministry of Finance) also have regulatory roles.

General organization and operating principles of trading markets are established by the AMF through its "Règlement général." The AMF's membership comprises various professional categories involved in the securities market such as exchange intermediaries, industrial and commercial firms, institutional investors and employees' representatives. The sixteen members of the AMF are appointed for a four-year period by decree of the ministry in charge of Economy and Finance.

The AMF is responsible for the proper working of regulated exchange markets. It supervises compliance with exchange rules and regulations by investment companies operating in France, and exchange compliance with the AMF's own rules. It has the power to impose sanctions on violators.

Nord Pool

Nord Pool was the first exchange to trade emissions allowances, listing EUA and CER contracts in February 2005 and June 2007 respectively, but it has not kept up with ECX and BlueNext in trading volume. It offers spot forward contracts for EUAs and CERs (where actual delivery of an allowance always occurs), in addition to its primary business, which is spot trading of electrical power among the Nordic countries.

As a regulated exchange, Nord Pool maintains a market surveillance operation, and reports formally to the Norwegian Kredittilsynet (Financial Supervisory Authority) and the NVE (the Norwegian energy regulator).

Market participants are required to report all non-exchange (OTC) trades, to disclose all inside information that is likely to have a price impact, and to refrain from trading while holding such information. The exchange provides a mechanism for disclosing price-sensitive information, called the Urgent Market Message (UMM), which can be sent any time day or night. Price manipulation is defined in Nord Pool rules and prohibited.

Nord Pool also operates a clearing house, which clears both exchange and OTC transactions and is registered with the Kredittilsynet.

European Energy Exchange (EEX)

The EEX, in Leipzig, Germany, operates a spot market for EUAs, in addition to a natural gas spot and futures market. It lists various emissions derivatives contracts, but trades lower volumes than London. Under German law, the exchange maintains a Market Surveillance Department (HUSt), which is autonomous and independent of the exchange itself. HUSt reports twice a month to the exchange supervisory authority, which is part of the Saxon Ministry for Economic Affairs and Labor. The supervisory authority conducts inspections and issues instructions to the exchange's management board, and may order HUSt to conduct investigations. HUSt's investigative authority extends beyond members of the exchange to anyone who may be involved in suspicious circumstances or grievances involving exchange trading or settlement.

HUSt also reports to the federal Financial Supervisory Authority (BaFin).

Because of the number of international trading participants, HUSt reports regularly to a range of foreign financial and energy regulators, including the U.S. CFTC.

End Notes

[1] P.L. 10 1-549, Title IV (November 15, 1990).

[2] Norway, a non-EU country, also has instituted a CO_2 trading system which is currently linked to the EU-ETS. Various other countries and a state-sponsored regional initiative located in the northeastern United States involving several states are developing or implementing mandatory cap-and-trade system programs. For a review of these emerging programs, along with other voluntary efforts, see CRS Report RL33812, *Climate Change: Action by States to Address Greenhouse Gas Emissions*, by Jonathan L. Ramseur.

[3] Other U.S. cap-and trade programs exist—most notably the nitrogen oxide program developed by EPA in the late 1990s.

[4] For more on the EU-ETS, see CRS Report RL34 150, *Climate Change and the EU Emissions Trading Scheme (ETS): Kyoto and Beyond*, by Larry Parker.

[5] Maria Mansanet-Bataller, Angel Pardo, and Enric Valor, "CO_2 Prices, Energy and Weather," 28 *The Energy Journal* 3 (2007), pp. 73-92. Powernext (a French energy exchange) has described CO2 prices as the cornerstone of relative energy prices for generating electricity. See Jean-Francois CONIL-LACOSTE, Chief Executive Officer, Powernext SA, *Market Based Mechanisms to Fight Climate Change* (2006).

[6] See, e.g., Senate Permanent Subcommittee on Investigations, "Excessive Speculation in the Natural Gas Market" (Staff Report), June 2007, 135 p.

[7] Point Carbon, *Carbon 2009: Emission Trading Coming Home* (March 2009), pp. 4-7. Point Carbon is a brokerage firm with its head office in Oslo, Norway.

[8] Kjetil Roine and Henrik Hasselknippe, "Emissions Trading in Europe," *Futures Industry*, May/June 2007. Figures are for 2006.

[9] The EUA swap market has been slow to develop for two reasons. First, there is no standard contract documentation— several private trade associations have competing versions, which make it difficult for a trader with many individual contracts to calculate his overall position in the market. Second, there is no standard, universally accepted price source to use to calculate a swap's value. Several investment banks, brokers, and markets offer competing prices and indices.

[10] In 2007, the secondary market in CERs amounted to 350 million metric tons, or \$8.3 billion. "Global Carbon Market Grows 80 Percent in 2007," *Business Wire*, January 18, 2008.

[11] European Commission, *EU Emissions Trading: An Open Scheme Promoting Global Innovation to Combat Climate Change*, Brussels, EU, 2005, p. 14.

[12] The exception, as always, is that trades must be reported to the national registries, which are linked electronically to a central EU facility. An ownership registry is necessary because all firms must surrender at the end of each year a quantity of allowances sufficient to cover their emissions (or else pay cash to cover the excess pollution).

[13] The website of the International Emissions Trading Association (IETA) has brief summaries of Polish, German, Dutch, and Czech laws on this point. See *http://www.ieta.org /ieta/www/pages*/index.php?IdSitePage=152, visited February 27, 2008.

[14] A derivatives clearing house guarantees that all contracts will be paid, even if an individual counterparty defaults. Clearing houses are a traditional feature of the futures exchanges: the guarantees they provide permit rapid trading to occur by eliminating the need for traders to assess the opposite party's creditworthiness. In recent years, clearing houses have begun to accept off-exchange, or over-the-counter (OTC), trades as well. LCH.Clearnet, a British clearing house, announced that it would provide clearing services for OTC emissions trades beginning on February 29, 2008. See LCH.Clearnet, Press Release, February 25, 2008.

[15] CFTC regulation, in many aspects, is modeled on Securities and Exchange Commission (SEC) regulation of the stock markets.

[16] For example, due to local legal provisions, LCH.Clearnet SA reports to French, Dutch, Belgium and Portuguese regulators as follows: Banque de France, Autorité des Marchés

Financiers, Comité des Etablissements de Crédit et des Entreprises d'Investissement, Commission Bancaire; Commission Bancaire, Financière et des Assurances, Banque Nationale de Belgique; and De Nederlandsche Bank, Autoriteit Financiële Markten. http://www.lchclearnet.com/ rules_and_regulations/sa/, visited February 26, 2008.

[17] Liz Bossley, *Emissions Trading and the City of London*, London, City of London, 2006, p. 31.

[18] Additionally, trading in EUAs began before the national registries were operational. As a result, all early trades were forward contracts, because immediate delivery was impossible.

[19] Alex Scott, "Europe's CO2 Permit Prices Dive; Further Turbulence is Expected," *Chemical Week*, vol. 168, May 17, 2006, p. 15.

[20] Statement of Per-Otto Wold, in U.S. Congress. Senate. Committee on Energy and Natural Resources, *EU Emissions*, Hearing, 110th Cong., 1st sess., March 26, 2007, p. 11.

[21] See, e.g., Robert Godby, "Market Power in Laboratory Emission Permit Markets," *Environmental and Resource Economics*, vol. 23, November 2002, p. 279; Akira Maeda, "The Emergence of Market Power in Emission Rights Markets: The Role of Initial Permit Distribution," *Journal of Regulatory Economics*, vol. 24, November 2003, p. 293; and Matti Liski and Juan-Pablo Montero, "A Note on Market Power in an Emission Permits Market with Banking," *Environmental and Resource Economics,* vol. 31, June 2005, p. 159.

[22] Christian Egenhofer, "The Making of the EU Emissions Trading Scheme," *European Management Journal*, vol. 25, December 2007, p. 455. This scenario is the opposite of what was suggested above: instead of trading ahead of the publication of price-sensitive information, the putative manipulators *fail to trade* on such information, maintaining an artificially high price. In another hypothetical scenario, large traders with inside information could prop up the price in the spot market to give them time to take short positions in the derivatives market that would become profitable when the spot price eventually fell.

[23] European Commission, *Directive 2009/29/EC of the European Parliament and of the Council of 23 April 2009 amending Directive 2003/8 7/EC so as to improve and extend the greenhouse gas emission allowance trading system of the Community* (Brussels, April 23, 2009), article 12(a).

[24] Clean Air Act Amendments of 1999, P.L. 101-549, Title IV. For a more detail discussion of the title, see Larry B. Parker, Robert D. Poling, and John L. Moore, "Clean Air Act Allowance Trading," 21 *Environmental Law* 2021-2068 (1991).

[25] EPA has renamed the ATS the Allowance Management System (AMS), but ATS remains the commonly used term and will be used in this chapter.

[26] 15 U.S.C. 79 et seq.

[27] 16 U.S.C. 791 et seq.

[28] For a comprehensive discussion of state and federal regulation, see Robert Poling, et. al., *Electricity: A New Regulatory Order?* Report for the Committee on Energy and Commerce, House of Representative (June 1991), committee print.

[29] P.L. 109-5 8.

[30] P.L. 95-617, 16 U.S.C. 2601.

[31] P.L. 95-620.

[32] Exempt Wholesale Generators may sell electricity only at wholesale. EWGs may be located anywhere, including foreign countries. Before enactment of EPACT05, utility generators were limited by the Public Utility Holding Company Act of 1935 (PUHCA) to operate within one state.

[33] In repealing PUHCA, EPACT05 provides that FERC and state regulatory bodies must be given access to utility books and records. Also, FERC is given approval authority over the acquisition of securities and the merger, sale, lease, or disposition of facilities under FERC's jurisdiction with a value in excess of $10 million. With respect to PURPA, EPACT05 repeals the PURPA mandatory purchase requirement for new contracts if FERC finds that a competitive electricity market exists and a qualifying facility has adequate access to wholesale markets. Among its provisions to address reliability, FERC is authorized to certify a national electric reliability organization (ERO) to enforce mandatory

reliability standards for the bulk power system. For more information on EPACT05, see CRS Report RL3 3248, *Energy Policy Act of 2005, P.L. 1 09-58: Electricity Provisions*, by Amy Abel.

[34] On October 3, 2001, the U.S. Supreme Court heard arguments in a case (*New York et al. v. Federal Energy Regulatory Commission*) that challenged FERC's authority to regulate transmission for retail sales if a utility unbundles transmission from other retail charges. In states that have opened their generation market to competition, unbundling occurs when customers are charged separately for generation, transmission, and distribution. Nine states, led by New York, filed suit, arguing that the Federal Power Act gives FERC jurisdiction over wholesale sales and interstate transmission and leaves all retail issues up to the state utility commissions. Enron in an amicus brief argued that FERC clearly has jurisdiction over all transmission and FERC is obligated to prevent transmission owners from discriminating against those wishing to use the transmission lines. On March 4, 2002, the U.S. Supreme Court ruled in favor of FERC and held that FERC has jurisdiction over transmission, including unbundled retail transactions.

[35] FERC Order 697, *Market-Based Rates for Wholesale Sales of Electric Energy, Capacity and Ancillary Services by Public Utilities*, Docket No. RM04-7-000, Final Rule (issued June 21, 2007).

[36] Code of Federal Regulations, Title 18, *Conservation of Power and Water Resources*, Part 101.

[37] Code of Federal Regulations, Title 18, *Conservation of Power and Water Resources*, Part 101. Allowance accounting is described under *General Instructions Number 21*.

[38] Treatment of emission allowances under the Federal income tax is spelled out in Rev. Rul. 92-16, *Internal Revenue bulletin*, No. 1992-12, March 23, 1992, p. 5 and Rev. Proc. 92-91, *Internal Revenue Bulletin*, No. 1992-46, November 16, 1992-13, p. 32-33. See also, Announcement 92-50, *Internal Revenue bulletin*, No. 1992-12, March 30, 1992, p. 32.

[39] For example, see Kenneth Rose, et. al., *Public Utility Implementation of The Clean Air Act's Allowance Trading Program*, National Regulatory Research Institute, May 1992.

[40] Elizabeth M Bailey, *Allowance Trading Activity and State Regulatory Rulings: Evidence from the U.S. Acid Rain Program*, MIT, March 1998, pp. 9-10.

[41] See Ken-Ichi Mizobuchi, *The Movements of PUC Regulation Effects in the SO2 Emission Allowance Market*, Kobe University, May 2004.

[42] "Official" here means that the transfer has been recorded by the ATS. The actual transfer of ownership may have occurred earlier. As noted earlier, parties are not required to notify the ATS of any transfer within a specific time period and may choose for some reason to delay informing the ATS of a transfer.

[43] For example, see the now terminated agreement AEP System Interim Allowance Agreement filed with the FERC on August 30, 1996 in Docket No ER96-2213-000 designated as Appalachian Power Company Supplement No. 9 to Rate Schedule FPC No. 20; Columbus Southern Power Company Supplement No. 3 to Rate Schedule FPC No. 30; Indiana Michigan Power Company Supplement No. 10 to Rate Schedule FPC No. 17; Kentucky Power Company Supplement No. 6 to Rate Schedule FPC No. 11; and, Ohio Power Company Supplement No. 9 to Rate Schedule FPC No. 23. Agreement terminated by FERC, effective January 1, 2002, in accordance with the mutual consent of the parties thereto.

[44] Peter Zaborowsky, *The Trailblazers of Emissions Trading*, Evolution Markets Inc. (April 23, 2002).

[45] Ibid. In December 2009, the monthly volume was estimated at 200,000 by Evolution Markets Inc. Evolution Markets Inc., *SO_2 Markets—December 2009* at *http://new.evomarkets.com* /index.php?page=Resources-Market_UpdatesMonthly_ Market_Updates.

[46] The first year an allowance may be used for compliance is called its "vintage." This situation can result in entities engaging in a "vintage swap." For example, a "vintage swap" may occur because one entity has excess allowances in the upcoming year (2008) but anticipates it will have insufficient allowances in 2009. Another entity may be in the opposite position

because of planning future emission reductions. The two entities agree to "swap" allowances to improve their allowance streams over these years.

[47] EPA Website: http://www.epa.gov/airmarkets/trading/buying.html.

[48] EMA Website: *http://www.environmentalmarkets.org/page.ww?section=About+Us&name=Company+Directory.*

[49] Comment of Matt Most, Emissions Market Association, as reported in *Platts Emissions Daily*, "Emissions market hails NYMEX move," February 15, 2005, p. 1.

[50] *Platts Emissions Daily*, "Emissions exchanges continue to grow SO_2, NOx futures markets," August 10, 2007, p. 1.

[51] Chicago Climate Futures Exchange, *Chicago Climate Futures Exchange to Launch Options market on Sulfur Financial Instrument Futures Contracts*, Chicago, April 5, 2007.

[52] An option that can only be exercised for a short, specified period of time just prior to its expiration, usually a single day. "American" options, however, may be exercised at any time before expiration.

[53] For current options market data, see http://www.ccfe.com/mktdata_ccfe/sfi_options.jsf.

[54] Details on the NYMEX options available at http://nymex.greenfutures.com/notices/ntm138.html.

[55] The CFTC has occasionally brought enforcement actions for fraud in the spot market, but these are rare. The legislative history does not suggest that Congress meant the CFTC to be a regulator of cash commodity markets.

[56] See CFTC approval of CCFE application for designation as a Contract Market: *Order of Designation: In the Matter of the Application of the Chicago Climate Futures Exchange, LLC for Designation as a Contract Market*, November 9, 2004.

[57] Comptroller of the Currency, Administrator of National Banks, Interpretive Letter #1040: Emissions Derivatives Proposal, September 15, 2005.

[58] Board of Governors, Federal Reserve System, JPMorgan Chase & C. New York, New York: Order Approving Notice to Engage in Activities Complementary to a Financial Activity, November 18, 2005.

[59] See table *Venues for the Trading of Exempt Commodities under the Commodity Exchange Act (CEA)*, available on the CFTC website at *http://www.cftc.gov/stellent/grou* ps/public/@newsroom/documents/file/ exemptcommoditiesvenues_091207.pdf.

[60] Evolution Markets LLC, "An Overview of Trading Activity and Structures in the U.S. Emissions Markets," NYMEX Emissions Futures Seminar, July 28, 2005.

[61] The Division of Market Oversight and The Division of Clearing and Intermediary Oversight, CFTC, DCM Designation Memorandum: *Application of Chicago Climate Futures Exchange, LLC ("CCFE") for Designation as a Contract Market pursuant to Sections 5 and 6(a) of the Commodity Exchange Act ("Act" or "CEA ") and Part 38 of Commission regulations*, November 3, 2004.

[62] U.S. territories account for the remaining 1%. Data from EPA, *Inventory of U.S. Greenhouse Gas Emissions and Sinks: 1990-2005,* April 15, 2007, p. ES-14.

[63] The EU addresses this issue by having the ETS cover only 45% of its emissions and no non-carbon dioxide emissions, as noted earlier. Still, it has 11,500 entities to oversee.

[64] Range based on EPA estimates for reducing emissions to 1990 levels by 2020 as required under S. 280. See EPA, *Analysis of The Climate Stewardship and Innovation Act of 2007*, July 16, 2007. For reference, a Phase 2 ETS allowance currently sells for about 20-25 euro. Data from the European Climate Exchange, http://www.ecxeurope.com/ default_flash.asp.

[65] The CCX created the ECX in 2005. Since 2006, CCX and ECX have been owned by Climate Exchange PLC, a publicly traded company listed on the AIM division of the London Stock Exchange.

[66] http://www.cantor.com/brokerage_services/co2e, visited February 29, 2008.

[67] "Nymex and Other Major Market Participants to Form the Green Exchange," *PR Newswire,* December 12, 2007.

[68] In a liquid market, there are many buyers and sellers, and traders have less reason to fear that their order to sell will cause the price to drop before the order can be filled (or vice versa). For large traders in particular, the impact of their trade orders on the market price is a major component of total transaction costs.

[69] With the exception, of course, of the rare occasions when price manipulation is suspected, as discussed below.

[70] See CRS Report RS21401, *Regulation of Energy Derivatives*, by Mark Jickling, for a discussion of issues associated with unregulated OTC trading.

[71] This chapter does not address fraud outside the secondary markets, such as falsification of emissions data. The EPA would be the appropriate agency to oversee and verify emissions measurement.

[72] OTC traders must be "eligible contract participants," defined in the Commodity Exchange Act as financial institutions, market professionals, corporations with a net worth greater than $1 million, or individuals with assets over $1 million, or income over $200,000.

[73] Both derivatives and securities are traded in private, less-regulated markets. Under federal securities law, companies may sell stocks or bonds to a limited number of institutional investors or wealthy individuals without having to register with the SEC.

[74] For a brief overview of the law, see CRS Report RS21 127, *Federal Securities Law: Insider Trading*, by Michael V. Seitzinger.

[75] There is a large theoretical literature, on both stock and futures markets, that argues that any restriction of insider trading is undesirable because it makes pricing less efficient.

[76] The "tippees" are likewise prohibited from trading on that information. See 17 CFR § 159. (The definition of "material nonpublic information" is essentially the same as in securities markets.)

[77] See U.S. Senate Permanent Subcommittee on Investigations, *Excessive Speculation in the Natural Gas Market*, staff report, June 2007. The argument that manipulation has occurred is not universally accepted. See Written Testimony of Acting CFTC Chairman Walter Lukken and Commissioner Michael Dunn before the Permanent Subcommittee on Investigations, Senate Committee on Homeland Security and Governmental Affairs, July 9, 2007 http://www.cftc.gov/ stellent/groups/public/@newsroom/documents/ speechandtestimony/opalukken-26.pdf, visited March 3, 2008.

[78] For an overview of the CFTC's anti-manipulation tools, see "Written Testimony of Acting CFTC Chairman Walter Lukken and Commissioner Michael Dunn before the Permanent Subcommittee on Investigations," July 9, 2007.

[79] All futures contracts are bilateral, and any given price movement causes equal but opposite gains and losses to the two sides.

[80] See CRS Report RS21401, *Regulation of Energy Derivatives*, by Mark Jickling.

[81] This was the case a few years ago with Nasdaq and Instinet, a trading system used only by Nasdaq intermediaries. The SEC responded by requiring that Nasdaq trading screens display price quotes from all electronic marketplaces handling Nasdaq shares.

[82] See, e.g., U.S. Government Accountability Office, *Commodity Futures Trading Commission: Trends in Energy Derivatives Markets Raise Questions About CFTC Oversight*, GAO-08-25, October 2007, 83 p.

[83] The Working Group includes the chairmen of the Fed, SEC, and CFTC, and the Secretary of the Treasury.

In: The Role Emissions in Climate Change Policy ISBN: 978-1-61728-986-6
Editors: Heather C. Perkins

Chapter 4

EMISSION ALLOWANCE ALLOCATION TO ELECTRICITY LOCAL DISTRIBUTION COMPANIES (LDCS): CONSIDERATIONS FOR POLICYMAKERS*

Jonathan L. Ramseur and Stan Mark Kaplan

SUMMARY

This chapter examines a major component of the emission allowance distribution strategy proposed in two major legislative vehicles under consideration in the 111th Congress: H.R. 2454 (Waxman- Markey), which passed the House on June 26, 2009; and S. 1733 (Kerry-Boxer), which was ordered reported by the Senate Committee on Environment and Public Works on November 5, 2009. Both bills (among other provisions) would establish a cap-and-trade program that would distribute—in the program's early years—a substantial portion of allowance value to electricity local distribution companies (LDCs) for the benefit of electricity consumers. Considering the magnitude of allowance value directed to LDCs, the potential consequences and challenges associated with this approach raise fundamental policy issues.

---

* This is an edited, reformatted and augmented edition of a United States Congressional Research Service publication, Report R41041, dated January 25, 2010.

Although an LDC allocation approach would be effective in mitigating cap-related electricity price increases at the household level, a closer examination of this strategy reveals potential concerns. Recent economic studies indicate that an electricity LDC allowance distribution strategy would alter the carbon price signal. The overall cost of the cap-and-trade program would increase if the price signal is channeled away from economic sectors with abundant low-cost abatement opportunities to economic sectors that have fewer such opportunities. In addition, if the price signal is dampened in one sector of the economy or for a particular subgroup of society, the signal may shift to other sectors or other groups, yielding unintended impacts. Several studies have estimated the efficiency loss. The range of estimates indicates that the magnitude of the efficiency loss due to LDC allocation would be uncertain. This uncertainty may pose a challenge for policymakers, when evaluating trade-offs between various allowance distribution strategies.

Regions that get electricity from more carbon-intensive sources are generally expected to face disproportionate impacts, and many have argued that an electricity LDC allocation approach could be designed to address these impacts. However, recent economic studies suggest that the differences across regions (for the average household) may be relatively modest. A question for policymakers is whether this disparity is worth addressing with an LDC allocation strategy, especially considering the implementation challenges involved. Robust data exist for the carbon intensity of electricity *at the point of generation*, but analogous data for the carbon intensity of electricity *distributed for consumption* are substantially less precise. This information gap would likely pose a significant challenge in terms of distributing emission allowances to LDCs, if policymakers seek to account for different carbon intensities of electricity. In addition, a range of stakeholders has expressed concern that the state Public Utility Commissions, which oversee LDCs, would implement the LDC allowance distribution requirements in different ways. Allowing different applications of allowance value (for the benefit of the electricity consumer) could lead to varied impacts across state lines. Some state agencies may allow LDCs to use allowance value to support efforts (unrelated to electricity bill rebates) that may not be allowed in other states.

Recent economic models indicate that most households would fare better with a lump-sum distribution. This is likely partly due to the expected efficiency loss associated with the LDC allotment. Another factor is that households in different regions use energy in different ways and consume fuels with different levels of carbon-intensity. The lump-sum distribution approach would likely provide households with more flexibility to offset the cost increases specific to their situation, whereas an LDC allocation would favor those impacted the most by electricity price increases.

INTRODUCTION

A diverse array of efforts to reduce greenhouse gas (GHG) [1] emissions are currently under way or being developed on the international, national, and sub-national level (e.g., individual state actions or regional partnerships). Recent legislative proposals have generally focused on using market-based approaches, such as a cap-and-trade system or emission fees ("carbon tax") system, to reduce GHG emissions.[2] So far, the cap-and-trade approach has generated far more congressional activity in terms of introduced bills and committee action. Indeed, a cap-and-trade proposal—H.R. 2454 (Waxman/Markey), which also includes numerous energy-related provisions—passed the House on June 26, 2009; the Senate Committee on Environment and Public Works ordered reported comparable legislation, S. 1733 (Kerry-Boxer), November 5, 2009.[3]

In a GHG cap-and-trade system, the cap is partitioned into emission allowances (or permits). Covered entities (e.g., power plants) would generally need to acquire and submit one allowance for each metric ton of carbon dioxide-equivalent (tCO_2-e) emitted in a previous year. Emission allowances would be the currency of a cap-and-trade program, potentially accounting, in aggregate, for tens or hundreds of billions of dollars per year.

In general, emission allowance value can be distributed in two ways (both of which may be utilized in the overall distribution strategy). Congress could allot free allowances to select parties (covered entities or non-covered entities), who would be able sell the allowances in the newly created emissions trading market. Alternatively, Congress could auction the allowances and redistribute the auction revenues to select parties or to support various policy objectives. Therefore, the distribution of emission allowance value is akin to the distribution of money. Although the emission allowance distribution strategy *would not* affect the environmental integrity of the cap-and-trade program (allowances would be limited, matching the emissions cap), the chosen approach would affect the distribution of costs associated with the emissions cap.

Therefore, one of the more controversial and challenging questions for policymakers is *how, to whom, and for what purpose* to distribute the emission allowances. The distribution of emission allowance value presents Congress with a series of trade-offs. The primary options involve (1) minimizing the overall program costs imposed on society;[4] (2) alleviating the costs borne by households or economic sectors; and (3) providing funding to support other policy objectives— deficit reduction, health care—which may or may not relate to climate change mitigation.

This chapter looks at various issues and concerns involving the distribution of emission allowances to electricity local distribution companies (hereafter LDCs), with a particular focus on this strategy in H.R. 2454 and S. 1733. The first section discusses relevant topics in the electric power industry, providing a context for the subsequent LDC allocation discussion. The second section examines the effectiveness and potential consequences of using the LDC allocation approach. The third section provides some conclusions.

ELECTRIC POWER SECTOR—AN OVERVIEW

The major components of the electricity sector include generation, transmission, distribution, and consumption (Figure 1).[5]

- Generation: Power plants produce electricity, using either combustible fuels such as coal, natural gas, and biomass; or non-combustible energy sources such as wind, solar energy, and nuclear fuel.
- Transmission: Power lines carry electricity from the power plant to demand centers. The higher the voltage of a transmission line the more power it can carry.
- Distribution: Step-down transformers reduce voltage so that distribution lines can deliver power to electricity customers.
- Consumption: Electricity is used by a variety of consumers—households, commercial establishments, and industrial facilities.

Regulation of Electric Power Sector

The regulatory structure for electricity generation is different than that of electricity distribution. In the United States, the price of *electricity generation* may be determined by a state regulatory body—often described as cost-of-service regulation—or the price may be subject to market forces—in "restructured" markets with competitive elements. In general, the regulatory structure varies by the type of facility and/or the state in which the electricity is generated. In 2007, the more traditional, price-regulated electric utilities generated approximately 60% of the total net electricity generated in the United States.[6]

In contrast to electricity generation, *electricity distribution* is generally regulated by a state agency, regardless of location. As discussed later, the different regulatory schemes (between generation and distribution) raise issues for policymakers regarding emission allowance distribution.

Electricity-Related Emissions and Reduction Opportunities

The electricity generating sector is of particular importance in a cap-and-trade system. Of the numerous activities that generate GHG emissions in the United States, the single greatest contributor is the electricity generation sector, accounting for approximately 34% of all U.S. GHG emissions in 2007.[7] Moreover, economic models of cap-and-trade proposals often indicate that the vast majority of the emission reductions achieved in the first few decades of the program would come from the electricity sector.[8] This is at least partially explained by the fact that—compared to other covered sectors—the electricity generating sector may offer more lower-cost opportunities for emission reductions: fuel substitution (e.g., coal to less carbon-intensive fuels), energy efficiency, and end-user conservation.

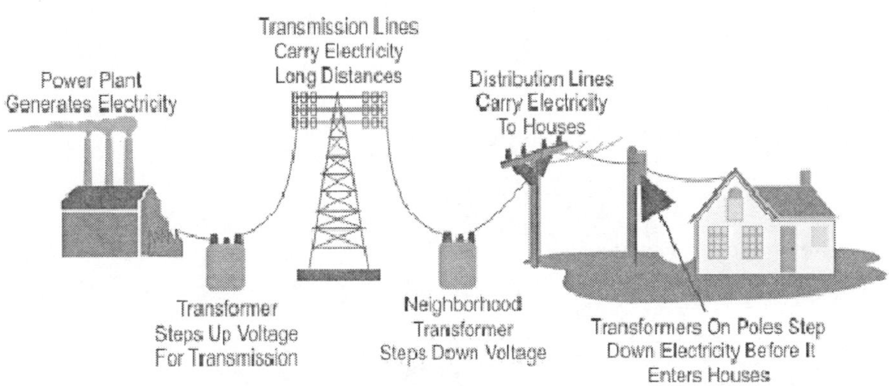

Source: Energy Information Administration, Electricity Basics 101, at http://www.eia.doe.gov/basics/ electricity_basics.html.

Figure 1. Illustration of the Electric Power Sector from Generation to Consumption

Emission Allowance Distribution—H.R. 2454

Expected cap-related price increases and their impacts have been a subject of intense debate during the development of cap-and-trade legislation in recent years. Recent cap-and-trade proposals have included design elements intended to address these price-related impacts.[9] Although policymakers have a variety of options available for this purpose,[10] one of the options that has received considerable interest is the distribution of emission allowances to electricity local distribution companies (LDCs).

Figure 2 indicates the emission allowance value distribution for H.R. 2454 in 2016 (S. 1733 has a similar, but not identical, allowance distribution strategy).[11] Both H.R. 2454 and S. 1733 would distribute—in the early years (2012-2029) of the cap-and-trade program—a substantial portion of allowance value to electricity local distribution companies (LDCs) for the benefit of electricity consumers. In the LDC context, "consumers" refers to households, government, businesses, and industrial facilities; that is, any customer that purchases electricity from a local utility. Some of the industrial end-users of electricity are also covered sources under the emissions cap.[12]

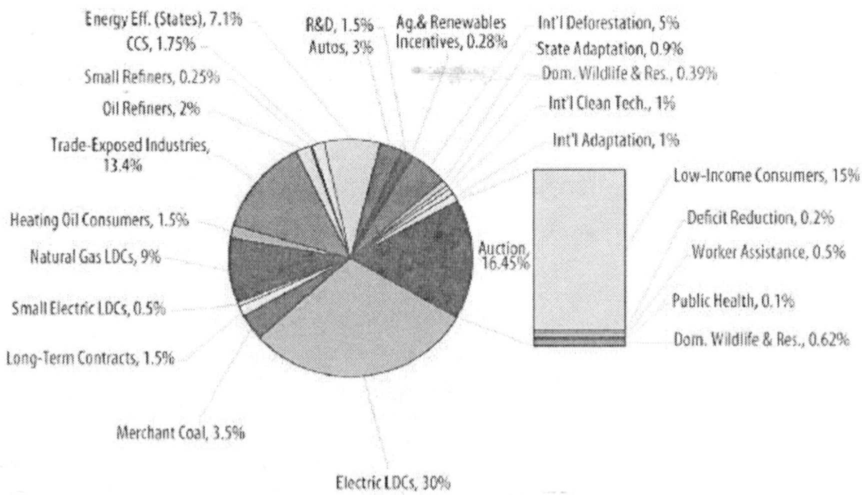

Source: CRS analysis of H.R. 2454, as passed by the House.

Notes: Percentages reflect the share of total allowances less those reserved for the Strategic Reserve. Under H.R. 2454, an additional 650 million allowances from future years would be auctioned in 2016 and the revenues used for deficit reduction. These allowances are not captured by the percentages above.

Figure 2. Emission Allowance Allocation Under H.R. 2454—2016

Both H.R. 2454 and S. 1733 would distribute LDC allowances based on a two-part formula. Fifty percent of the allowances would be distributed to LDCs based on deliveries of electricity. The remaining 50% would be distributed to LDCs based on the carbon intensity of their delivered electricity in proportion to other LDCs. This distribution formula would mean that, all else being equal, an LDC that delivers a higher percentage of coal-fired electricity would receive a greater proportion of the allowances than LDCs that deliver electricity generated from other sources. This latter component of the formula would seek to address the higher electricity prices that consumers would face in locales with a greater proportion of coal-fired electricity.

ALLOCATION TO ELECTRICITY LDCS: EFFECTIVENESS AND POTENTIAL CONSEQUENCES

In general, the primary arguments voiced in support of the LDC allowance distribution strategy are (1) to help mitigate the expected electricity price increases (particularly for households) that would occur from an emissions cap; and (2) to help address potential regional differences in electricity prices that would occur from an emissions cap. This section discusses the effectiveness and possible consequences of using an LDC allocation approach to further these objectives. In so doing, this section provides select comparisons between an LDC allocation approach and a lump- sum distribution of allowance value to households.

Mitigating Electricity Price Impacts

A GHG emissions control program, such as cap-and-trade, places a price on carbon. The carbon price will be reflected in higher energy prices (e.g., fossil fuels) as well as other materials and services, particularly electricity, that are generated through energy use. As Table 1 indicates, the Environmental Protection Agency (EPA) estimates (within the framework of H.R. 2454) that the national *average* price of electricity would increase by a larger percentage compared to the prices of natural gas and oil. Note that these price increases do not account for distribution of allowance value or behavioral responses.

Table 1. EPA Estimate of Select Price Increases under H.R. 2454 Compared to a "Business-as-Usual" Scenario Figures Represent Percentage Increases without Distribution of Emission Allowance Value

	2015	2020	2025
Electricity	10.7	12.7	14.0
Natural gas	7.4	8.5	8.6
Petroleum	3.2	4.0	4.7

Source: Prepared by CRS with data from EPA, *Analysis of the American Clean Energy and Security Act of 2009, H.R. 2454 in the 111th Congress* (June 2009), Data Annex, ADAGE Scenario 6.

Notes: The percentage increases are dependent on the estimated baseline prices. The prices of natural gas and oil have fluctuated dramatically in recent years. If the baseline prices of either were to increase/decrease, the percentage impact of the emissions cap on the fuel price would decrease/increase, affecting the relative comparison to electricity price increases. The 2015 baseline prices used by EPA included \$0.085/kilowatt-hour for electricity; \$8.14/million Btu for natural gas (approximately \$8.36 per thousand cubic feet); and \$23.32/million Btu for petroleum (approximately \$135 per barrel).

These expected price increases would impact energy consumers. Absent the redistribution of allowance value to help offset increased energy costs, household consumers would likely bear a substantial portion of the costs imposed by a cap-and-trade program. This would be due to the ability of covered entities (e.g., power plants, petroleum producers/importers, large industrial facilities) and non-covered entities to pass on (1) the costs incurred from complying with the program; and/or (2) higher energy costs associated with business operations.[13] Figure 3 compares the cap-related impacts on households (as a percentage of household income) before allowance value distribution with an LDC allocation approach. As Figure 3 illustrates, households at all income levels would be worse off without allowance value distribution, as indicated by the negative impacts to their income; low-income households would bear a greater burden (as a percentage of their income) compared to higher-income levels—often described as a regressive outcome. In contrast, Figure 3 indicates all income levels would be better off with an emission allowance allocation to electricity LDCs. The most dramatic contrast would occur at the lowest income households, which would actually see benefits rather than costs under this scenario.

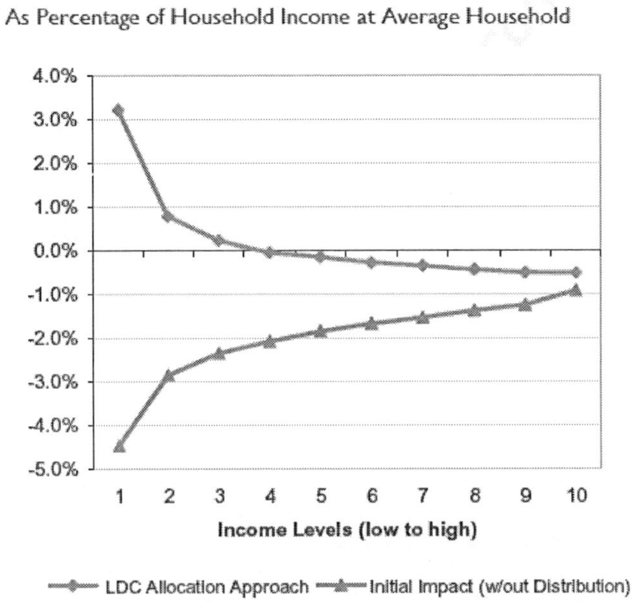

Source: Prepared by CRS; data from Rich Sweeny and Dallas Burtraw, *The Effects on Households of Allocation to Electricity Local Distribution Companies* (May 19, 2009).

Notes: These estimates are based on a hypothetical cap-and-trade program that would yield an emission allowance price of approximately $21 in 2015. In the "LDC Allocation Approach" depicted above, 30% of the allowance value would go to LDCs; the remaining 70% would be distributed to households in a lump-sum fashion. For more information on the model used to develop the estimates, see Dallas Burtraw, et al., *The Incidence of U.S. Climate Policy: Alternative Uses of the Revenue from a Cap-and-Trade Program* (June 2009), Resources for the Future Discussion Paper 09-17-Rev.

Figure 3. Estimated Impacts on Households of Providing Allowances at No Cost to Electricity Local Distribution Companies

Potential Concerns

Although the above comparison indicates that an LDC allocation approach could be effective in mitigating cap-related impacts at the household level, a closer examination of this strategy reveals potential concerns associated with its use. Moreover, would an alternative allocation strategy be more effective at mitigating cap-related impacts on households? These issues are discussed below.

Altered Price Signal and Its Effects

A primary concern shared by some observers is that an LDC allowance distribution strategy would alter the carbon price signal inherent to the cap-and-trade system. The consequences of an altered price signal may be two-fold: (1) a less economically efficient program; and (2) an unintended and undesirable shift in the price signal.

Although cap-related price increases may impose substantial burdens on some economic sectors or individuals, price increases for carbon-intensive products, such as electricity, are a critical design feature of a market-based GHG emission reduction program. As stated by the Director of the Congressional Budget Office (CBO):

> The price increases would be essential to the success of a cap-and-trade program because they would be the most important mechanism through which businesses and households would be encouraged to make economically motivated changes in investment and consumption that reduced CO_2 emissions.[14]

If the price signal is channeled away from economic sectors with more GHG abatement opportunities to economic sectors that have fewer opportunities, the overall cost of the cap-and-trade program would increase. As mentioned earlier, the electricity sector would offer relatively more emission reduction opportunities from fuel substitution, end-user conservation, and energy efficiency options on the producer and consumer ends. The Director of the CBO recently described the effects of using emission allowance value to hold electricity bills steady:

> Muting the increase in electricity prices would increase the overall cost of the policy because it would reduce households' incentives to undertake measures to reduce their electricity consumption, such as choosing more efficient appliances or turning down their thermostats.[15]

A few studies have provided estimates of the efficiency loss (measured as increases in emission allowance prices) that would occur by providing some portion of allowances to LDCs. The estimates vary, because both the allowance distribution strategies and the underlying assumptions in the studies' models differ:

- A 2008 study from Resources for the Future (RFF) compared a 100% "lump-sum" distribution to households with a 100% no-cost distribution to LDCs, and found that the LDC allocation would raise the allowance price by approximately 13%. [16]
- A May 2009 RFF study compared a 100% "lump-sum" distribution to households with a distribution strategy of 70% to households and 30% to electricity LDCs, and found the LDC allotment approach would raise the allowance price by approximately 29%.[17]
- A June 2009 RFF study, based on the May study's basic structure, altered the model's behavioral assumptions for industrial and commercial customers responses to price increases. This change in the model allowed for a more efficient outcome compared to the May results: the allowance price increased by 9%.
- A 2009 interagency report (prepared in response to a letter from five Senators) stated (in a footnote) that in June 2009 EPA's analysis of H.R. 2454, the results "imply that LDC allocations would raise allowance prices by only 2 percent."[18] However, this interpretation of EPA's analysis is debatable, because EPA did not specifically examine the impacts of LDC allocations in its allowance price analyses. Although the agency's sensitivity analysis included a scenario without LDC allocations, this scenario differed in other ways that cloud the effects of adding/removing LDC allocation.

The range of above estimates indicates that the magnitude of the efficiency loss due to LDC allocation would be uncertain. This uncertainty may pose a challenge for policymakers, when evaluating the trade-offs between various allowance distribution strategies.

In addition to the possible efficiency loss (reflected in higher allowance prices), if the price signal is dampened in one sector of the economy or for a particular subgroup of society, the signal will move to other sectors or other groups. This could yield unintended, disproportionate impacts. During the testimony cited above, the CBO Director described this possibility:

> As a result [of muting the increase in electricity prices], the burden of meeting the cap would fall more heavily on other sectors, and that additional burden would be reflected in higher prices for other goods and services that

households purchase. (For example, the price of gasoline would probably increase more than would otherwise be the case.)[19]

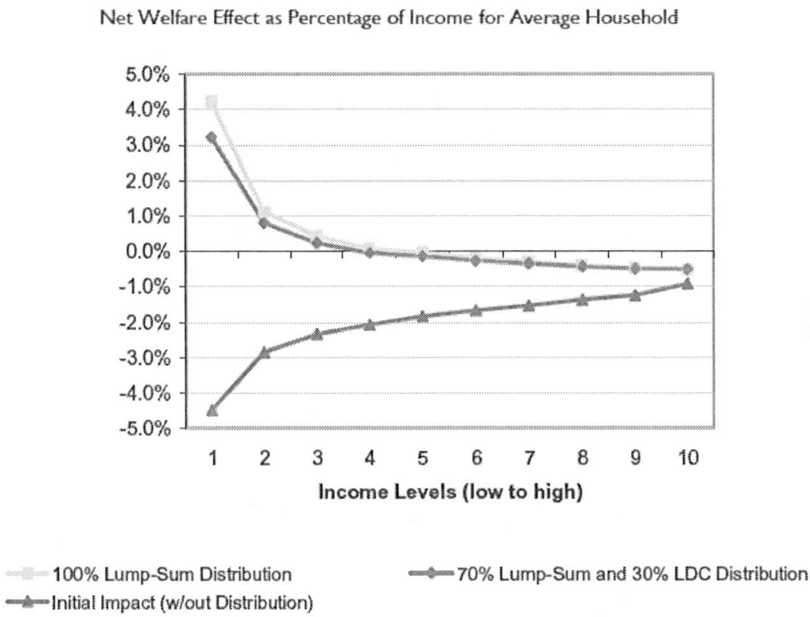

Source: Prepared by CRS; data from Rich Sweeny and Dallas Burtraw, *The Effects on Households of Allocation to Electricity Local Distribution Companies* (May 19, 2009).

Notes: These estimates are based on a hypothetical cap-and-trade program that would yield an emission allowance price of approximately $21 in 2015. For more information on the model used to develop the estimates, see Dallas Burtraw, et al., *The Incidence of U.S. Climate Policy: Alternative Uses of the Revenue from a Capand-Trade Program* (June 2009), Resources for the Future Discussion Paper 09-17-Rev.

Figure 4. Estimate of Cap-Related Impacts on Households for Different Allowance Distribution Approaches

A price signal shift from electricity to gasoline could be more detrimental to households. According to the Bureau of Labor Statistics' Consumer Expenditure Survey, the average household consumer spends 2.7% of his income on electricity and 5.4% on gasoline and motor oil.[20] For the average consumer, this equates to approximately $1,363 on electricity and $2,726 on gasoline. Because of these magnitude differences, a percentage increase in gasoline expenses may impose a greater loss to the average consumer than a percentage increase in electricity expenses. Another factor that could affect these impacts is the responsiveness of consumers to changes in electricity prices and gasoline prices. Consumers' sensitivities to either price increase may vary.[21]

LDC Allocation Versus "Lump-Sum" Distribution to Households

Although an allotment of allowances to electricity LDCs would help households cope with increased electricity prices (Figure 3), the 2009 RFF economic model (both May and June, cited above) indicates that an alternative distribution approach would generally be more beneficial to households. The RFF study compared an emission allowance distribution strategy that allots 100% of allowance value to households ("lump-sum") and one that allots 70% to households and 30% to electricity LDCs.[22] Figure 4 illustrates the *national average* impacts of the two distribution strategies for each household income level. The figure depicts the impacts as a percentage change in average household income. The figure shows that, when looking at the national average estimates, all income groups (at the national average) would be slightly better off under the 100% household lump-sum distribution approach. The reason for the difference relates to projected efficiency loss and the price signal shift (discussed above) that would ensue (to some degree) from the LDC allocation.

However, in a handful of regions, the highest income level households fared better with the LDC allocation approach. This likely relates to the increased profits projected in the electricity sector and the fact that the majority of shareholders are in upper-income households. Although H.R. 2454 and S. 1733 would provide emission allowances to electricity LDCs "exclusively for the benefit" of electricity consumers, some research indicates that the allowance value would effectively benefit electricity generating facilities and their shareholders.[23] As explained in the research, free allocation to electricity LDCs would depress the price signal in the electricity sector, which could lead to increased electricity consumption. The research estimates that generators in competitive regions would see annual profits increase by approximately $2.5 billion; generators in state-regulated markets may also see increased profits if lower prices lead to increased sales.[24]

Addressing Regional Differences

Many expect geography to play a key role in determining cap-related impacts, largely due to different energy uses across regions. In particular, energy sources used to generate electricity vary across regions, yielding different levels of GHG emissions per unit of electricity supplied, a measure known as carbon intensity of electricity: Coal has almost twice the carbon intensity (per unit of energy) of natural gas; the carbon intensity of electricity from nuclear, hydropower, and renewable energy sources is effectively zero.[25] Regions that get electricity from more carbon- intensive sources are generally expected to face disproportionate impacts, compared to those that rely on less-carbon intensive fuels. Many have argued emission allowance value should be distributed in a manner that would address these regional differences. This section assesses this underlying expectation, using data from two economic studies that modeled household impacts across various regions under a hypothetical GHG reduction program. Next, this section discusses potential implementation challenges in addressing different cap-related impacts across regions. Lastly, this section compares how households fare in different regions using an LDC allocation approach versus a lump-sum distribution strategy.

Expected Regional Differences in Electricity Prices: Cause for Concern?

Two recent economic studies—one affiliated with the National Bureau of Economic Research (NBER) and one affiliated with RFF—have examined household economic impacts by region under a hypothetical GHG reduction program.[26] Both studies' estimates illustrate the relative *gross* impacts on households under a GHG emission control program, without redistribution of allowance value to any entities. Thus, these estimates provide a rough approximation of impacts across different regions. The studies found similar results for the *average* household:[27]

- The NBER study estimated increased costs (as a percentage of annual income for the average household) would range from 1.5% in New England to 1.9% in the East-South-Central region. The study examined the effects of a \$15 (per metric ton of CO_2) carbon tax in 2003 (using 2005 dollars).
- The RFF study estimated increased costs (as a percentage of annual income for the average household) would range from 1.3% in California to 1.6% in the Ohio Valley (Figure 5). The study examined

the impacts of an emission allowance price of $20.91 (per metric ton of CO2) in 2015 (in 2006 dollars).

To put these percentage ranges in context using RFF data,[28] the average house in the Ohio Valley region would see costs rise by $54 more than the average household in California. As with cap-and-trade modeling results in general, such estimates should be viewed with attentive skepticism:[29] these estimates are linked to the assumptions and constraints associated with the model, in particular, the assumed allowance price of $20.91. If a higher allowance price were employed, the magnitude of cost differences between regions would show a corresponding increase. Regardless, the estimated differences for the average household between regions are (arguably) modest.

However, the RFF study found that regional differences are more pronounced at different income levels. (The RFF study is unique in that it provided cost impact estimates to households by geographic region *and by income level.*) Figure 5 shows the lowest-income household in California would (on average) experience costs of 4.0% of its income; the corresponding household in the Ohio Valley region would bear costs of 5.5% of its income. In dollar terms, this would be a difference of $100. Considering the potential implementation challenges involved (discussed below), a question for policymakers is whether this disparity is worth addressing through an emission allowance distribution strategy that accounts for regional differences (or whether the disparity on the basis of income is intrinsically more important than the disparity on the basis of geography).

Potential Implementation Challenges

If policymakers want to address the expected electricity price disparities across different regions, legislation would need to consider the different carbon intensities of electricity, which varies based on the energy resource used to generate the electricity. As the geographic area in question becomes more refined—moving from a region to a state, or a state to an LDC—determining the carbon intensity of electricity becomes more difficult.

As discussed below, robust data exist for the carbon intensity of electricity *at the point of generation*. However, analogous data for the carbon intensity of electricity distributed for consumption are substantially less precise. This information gap would likely pose a significant challenge in terms of distributing emission allowances to LDCs based on their carbon intensity of electricity.

Carbon Intensity of Electricity Data at Generation

The carbon intensity of electricity (i.e., the mix of fuels used to *generate* electricity) at the level of the individual power plant is well documented.[30] The data are collected by the Energy Information Administration (EIA), an arm of the Department of Energy. EIA collects data on power generation by fuel type for all generating units with a capacity of one megawatt or greater that are connected to the electric power grid. Power plant operators are legally obligated to timely and accurately file the data with EIA. The data are collected monthly for large generating units and annually for smaller units, using the EIA-923 survey instrument.[31]

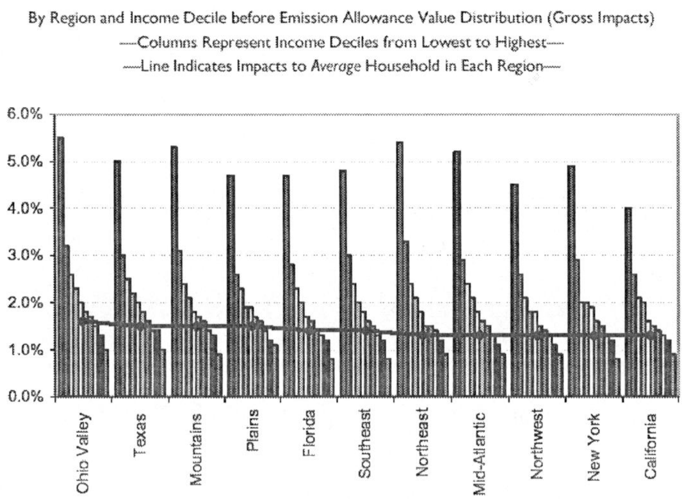

Source: Prepared by CRS, based on data from Dallas Burtraw, *et al.*, *The Incidence of U.S. Climate Policy: Alternative Uses of the Revenue from a Cap-and-Trade Program* (June 2009), Resources for the Future (RFF) Discussion Paper 09-17-Rev.

Notes: The model assumed a CO_2 emission control program was enacted in 2009; the above impacts correspond to 2015 with an assumed emission allowance price of \$20.91/metric ton of CO_2 (in 2006 dollars). The impacts illustrate relative differences between income groups that would occur if a price were imposed on CO_2 emissions. These percentages do not account for any offsetting income from emission allowance (or auction revenue) distribution. The percentages illustrate the relative gross impacts to households under a capand-trade program without redistribution of allowance value. The RFF study's multi-state regions include the following states: (1) Ohio Valley: Illinois, Indiana, Kentucky, Michigan, Missouri, Ohio, West Virginia, Wisconsin; (2) Mountains: Arizona, Colorado, Nevada; (3) Plains: Kansas, Minnesota, Nebraska, Oklahoma, South Dakota; (4) Southeast: Alabama, Arkansas, District of Columbia, Georgia, Louisiana, Mississippi, North Carolina, South Carolina, Tennessee, Virginia; (5) Northeast: Connecticut, Maine, Massachusetts, New Hampshire, Rhode Island; (6)

Mid-Atlantic: Delaware, Maryland, New Jersey, Pennsylvania; and (7) Northwest: Idaho, Montana, Oregon, Utah, Washington. For reasons discussed in the data source, seven states are not represented: Alaska, Hawaii, Iowa, New Mexico, North Dakota, Vermont, and Wyoming.

Figure 5. Estimates of Increased Cap-Related Household Costs as a Percentage of Average Household Income

Carbon Intensity of Electricity Data at Distribution

In many and probably most cases, an LDC does not have a complete picture of the energy sources—fossil fuels, nuclear, renewables, etc.—used to produce its power supply. Depending on the distribution utility's circumstances, it may have almost no relevant information. This is because of the following factors:

- Neither state governments nor the federal government collect data needed to accurately and routinely tie electricity deliveries to the generating source. Because of the complexity of power market transactions, in which a block of power can change hands several times before final delivery and the mix of fuels is constantly changing over time, it is unlikely that such a data collection process could be designed and implemented.
- Many municipal and cooperative utilities function entirely as distribution utilities. They buy power under a combination of long-term and short-term agreements with generating companies, and may also buy power daily off the grid at spot prices. These utilities do not have (or normally need) information on the source of the power.[32] In particular, a distribution utility making a spot market purchase at the market price does not know what mix of fuels is actually being used to produce the electricity. It may make the purchase from a power marketing company, not from a generator, which further removes the buyer from information about the original source of the electricity.
- Even utilities that generate their own electricity will frequently buy and sell large quantities of electricity, either under bilateral contracts or on the daily spot market. As discussed above, the fuel used to generate power is often not apparent in bulk power transactions. Therefore a large part of the power ultimately delivered to customers may not have a clearly defined source, either in terms of the generating units or fuels used to produce the power.

There may be some cases where a utility (1) generates its own power; and/or (2) buys and sells very little or no power off its own system. In those cases the utility would be able to reasonably estimate the fuel mix behind its power sales. However, these cases are in the minority.

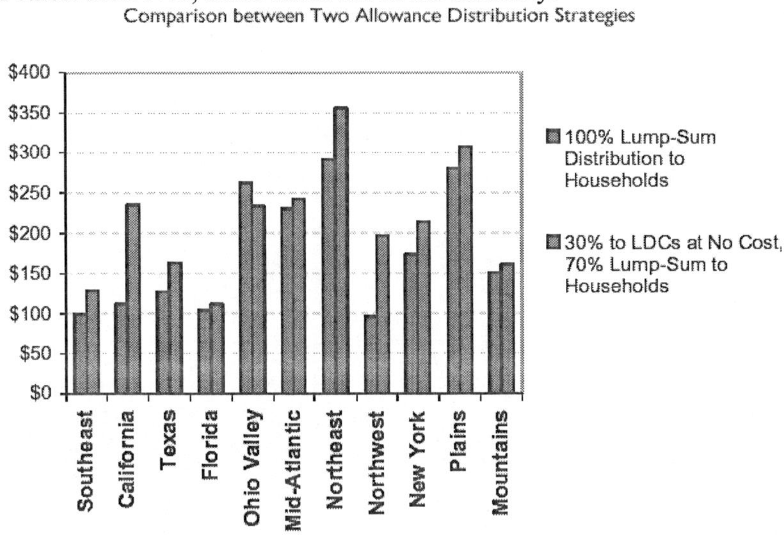

Source: Prepared by CRS; data from Rich Sweeny and Dallas Burtraw, *The Effects on Households of Allocation to Electricity Local Distribution Companies* (May 19, 2009).

Notes: These estimates are based on a hypothetical cap-and-trade program that would yield an emission allowance price of approximately $21 in 2015. For more information on the model used to develop the estimates, see Dallas Burtraw, et al., *The Incidence of U.S. Climate Policy: Alternative Uses of the Revenue from a Cap-and-Trade Program* (June 2009), Resources for the Future Discussion Paper 09-17-Rev.

The RFF study's multi-state regions include the following states: (1) Ohio Valley: Illinois, Indiana, Kentucky, Michigan, Missouri, Ohio, West Virginia, Wisconsin; (2) Mountains: Arizona, Colorado, Nevada; (3) Plains: Kansas, Minnesota, Nebraska, Oklahoma, South Dakota; (4) Southeast: Alabama, Arkansas, District of Columbia, Georgia, Louisiana, Mississippi, North Carolina, South Carolina, Tennessee, Virginia; (5) Northeast: Connecticut, Maine, Massachusetts, New Hampshire, Rhode Island; (6) Mid-Atlantic: Delaware, Maryland, New Jersey, Pennsylvania; and (7) Northwest: Idaho, Montana, Oregon, Utah, Washington. For reasons discussed in the data source, seven states are not represented: Alaska, Hawaii, Iowa, New Mexico, North Dakota, Vermont, and Wyoming.

Figure 6. Estimated Annual Cap-Related Losses to Average Households by Region

Applying Regional Estimates of Carbon Intensity of Electricity

Some groups have suggested using regional estimates in lieu of LDC-specific data.[33] For example, EPA has developed output emission rates for subregions in the agency's Emissions and Generation Resource Integrated Database (eGRID).[34] However, these estimates are not the same as carbon intensity of electricity *consumed*, because electricity may be generated in one region and exported for consumption to another region. Therefore, if policymakers use data that measures carbon intensity of electricity generated to distribute allowance value, regions that are net exporters of high-carbon electricity may be overcompensated at the expense of regions that are net importers of high-carbon electricity.

Moreover, distributing allowance value to LDCs based on regional data may overcompensate some LDCs at the expense of others. Regions (at least eGRID) comprise multiple states in many cases. Within a particular region (or state), the LDCs' carbon intensities of electricity likely span a wide range. For example, if one LDC were to purchase power solely from a nuclear power plant, while an LDC in a neighboring area purchased solely from a coal-fired generator, the two LDCs' carbon intensities of electricity would vary dramatically.

LDC Allocation Versus "Lump-Sum" Distribution to Households

Although proponents of the electricity LDC allocation strategy maintain that this approach would address different cap-related price impacts across regions,[35] the 2009 RFF study (discussed above and cited in **Figure 4**) estimated that the *average household* in most regions would fare better with a 100% lump-sum distribution than under a strategy combining a lump-sum (70%) with an allotment (30%) to electricity LDCs (hereafter referred to as the 70/30 strategy). In some regions, the percentage difference between costs for an average household would be substantial (Figure 6): the cost difference would be over 100% in the California and Northwest regions. The only region that fared better under the 70/30 strategy was the Ohio Valley, which is comprised of eight states.[36]

Arguably, the differences in dollars (or magnitude) are relatively small in this study, but these estimates are linked to the estimated allowance price of $21. If the modelers increased the estimated allowance price, the estimated difference in dollars would correspondingly increase.[37]

The explanation for the varied impacts across regions between the two allocation strategies may relate to (1) the efficiency cost from the LDC allocation (discussed above); and (2) the varied energy costs (and indirect costs) that households in different regions would face (Figure 7). Households in different regions use energy in different ways and consume fuels with different levels of carbon-intensity. In some regions for example, electricity would account for a much larger proportion of cap-related costs. As Figure 7 indicates, the electricity cost impact for the average household in the Ohio Valley would be roughly four times that of the average household in California or the Northwest region. Other regions would experience greater effects from fuel oil (e.g., Northeast) or natural gas cost increases (e.g., Plains). Note further the different projections of indirect cost impacts, which would include spending on goods and services that would be more expensive in a cap-and-trade system. The lump-sum distribution approach would likely provide households with more flexibility to offset the cost increases specific to their situation, whereas an LDC allocation would favor those impacted the most by electricity price increases.

Varied Implementation by State Agencies

Some observers have expressed concern that the state Public Utility Commissions or local agencies, which regulate electricity LDCs and would oversee allowance value applications, would provide different interpretations of the legislative provisions (in H.R. 2454 and S. 1733) pertinent to electricity LDCs.[38] For example, both bills state:

> To the extent an electricity local distribution company uses the value of emission allowances distributed under this subsection to provide rebates, it shall, to the maximum extent practicable, provide such rebates with regard to the fixed portion of ratepayers' bills or as a fixed credit or rebate on electricity bills.[39]

The manner in which the relevant agencies interpret such provisions could lead to varied impacts across state lines. For instance, some state agencies may allow LDCs to use allowance value for purposes (unrelated to electricity bill rebates) that may be questionable in other states. As noted in a recent study from the Massachusetts Institute of Technology,

> Different PUCs in different regions of the country may pursue different strategies for using this allowance revenue with very different distributional implications.[40]

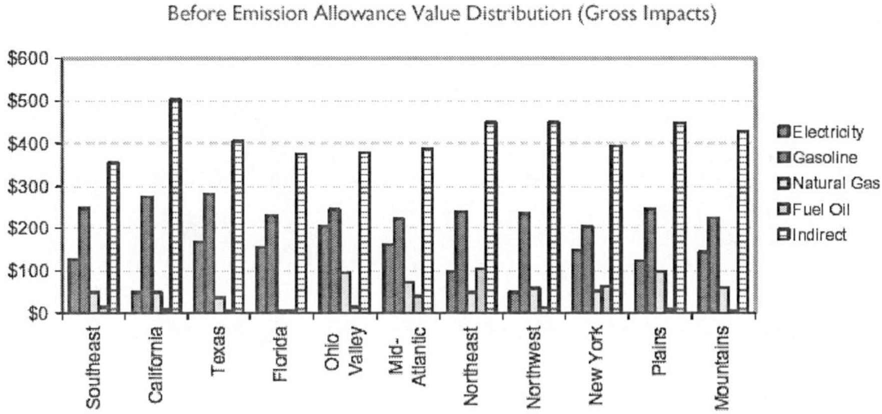

Source: Prepared by CRS with data from Dallas Burtraw, *et al.*, *The Incidence of U.S. Climate Policy: Alternative Uses of the Revenue from a Cap-and-Trade Program* (June 2009), Resources for the Future (RFF) Discussion Paper 09-17-Rev.

Notes: The model assumed a CO_2 emission control program was enacted in 2009; the above impacts correspond to 2015 with an assumed emission allowance price of $20.91/metric ton of CO_2 (in 2006 dollars). The impacts illustrate relative differences between regions that would occur if a price were imposed on CO_2 emissions. The estimates illustrate the relative gross impacts to households under a cap-and-trade program without redistribution of allowance value—a scenario that is unrealistic.

The RFF study's multi-state regions include the following states: (1) Ohio Valley: Illinois, Indiana, Kentucky, Michigan, Missouri, Ohio, West Virginia, Wisconsin; (2) Mountains: Arizona, Colorado, Nevada; (3) Plains: Kansas, Minnesota, Nebraska, Oklahoma, South Dakota; (4) Southeast: Alabama, Arkansas, District of Columbia, Georgia, Louisiana, Mississippi, North Carolina, South Carolina, Tennessee, Virginia; (5) Northeast: Connecticut, Maine, Massachusetts, New Hampshire, Rhode Island; (6) Mid-Atlantic: Delaware, Maryland, New Jersey, Pennsylvania; and (7) Northwest: Idaho, Montana, Oregon, Utah, Washington. For reasons discussed in the data source, seven states are not represented: Alaska, Hawaii, Iowa, New Mexico, North Dakota, Vermont, and Wyoming.

Figure 7. Estimated Annual Cap-Related Household Impacts by Category in Different Regions

Indeed, the National Association of Regulatory Utility Commissioners (NARUC) has highlighted this issue. In an October 28, 2009, letter to Chairman Boxer and Ranking Member Inhofe of the Senate Environment and Public Works Committee, NARUC stated the LDC "provisions are problematic because they foster uncertainty and potential litigation."[41] In particular, NARUC cited the legislative text requiring that allowance benefits be distributed "ratably" (among ratepayer classes) and "equitably" (among individual ratepayers):

> Just who will determine what "ratably" and "equitably" mean, and how will that impact the decision to issue flat rebates when the industrial and residential consumers demonstrate that their power bills have increased?[42]

Policymakers could address these concerns by providing explicit stipulations or limitations regarding the uses of emission allowance value. However, this approach would likely remove some flexibility, which may be ultimately beneficial to consumers in some parts of the country.

CONCLUSIONS

In recent years, cap-and-trade proposals have allocated increasing percentages of emission allowance value toward energy consumers, often corresponding with decreasing allotments to covered entities (see Table A-1). In both H.R. 2454 and S. 1733, electricity LDCs would receive a substantial portion of allowance value to be used for the exclusive benefit of electricity consumers. Considering the magnitude of allowance value involved (tens of billions of dollars per year), the potential consequences and challenges associated with this approach are receiving increasing attention.

A primary rationale for this allocation is to help mitigate the expected electricity price increases imposed by an emissions cap. Although an LDC allocation approach would be effective in mitigating cap-related impacts at the household level, a closer examination of this strategy reveals potential concerns. In short, this approach may yield consequences that contradict this policy goal. Depending on how the allowance value is applied, which would likely vary state by state, the allowances may dampen the carbon price signal in the electricity sector and channel the carbon price to other sectors. Because the electricity sector provides a large pool of relatively low-cost emission reduction opportunities, a price signal shift would likely be an economically inefficient outcome: i.e., pose a greater cost on the entire economy. Moreover, a price signal shift may increase the prices of other goods (e.g., gasoline) that may account for a greater percentage of annual households' expenses than electricity.

Several studies estimated the efficiency loss that would be expected to occur by providing some portion of allowances to LDCs. The range of estimates from the studies indicates that the magnitude of the efficiency loss due to LDC allocation would be uncertain. This uncertainty may pose a challenge for policymakers, when evaluating the trade-offs between various allowance distribution strategies.

Another stated purpose of the LDC allocation is to address potential regional differences in electricity prices that an emissions cap would pose. Although many may expect these differences to be substantial under a cap-and-trade system, two recent economic studies suggest that the differences across regions (for the average household) would be relatively modest. (However, the magnitude of the differences would be tied to the emission allowance price—assumed to be approximately $1 5-$20 in the studies—whose estimates are a subject of intense debate.) A question for policymakers is whether this disparity is worth addressing with an LDC allocation strategy, especially considering the implementation challenges involved.

Implementation would be problematic, because the information needed—carbon intensity of electricity delivered at the LDC level—is generally not available. Policymakers could utilize regional estimates in lieu of LDC-specific data, but this approach would likely overcompensate regions that are net exporters of high-carbon electricity at the expense of regions that are net importers of high-carbon electricity. Moreover, a regional approach may overcompensate LDCs at the expense of other LDCs, because LDCs' carbon intensities of electricity sometimes span a wide range within a particular region.

In addition, a range of stakeholders has expressed concern that the state Public Utility Commissions, which oversee LDCs, would implement the LDC allowance distribution requirements in different ways. Allowing different applications of allowance value (for the benefit of the electricity consumer) could lead to varied impacts across state lines. Some state agencies may allow LDCs to use allowance value to support efforts (unrelated to electricity bill rebates) that may not be allowed in other states.

Recent economic studies indicate that most households would fare better with a lump-sum distribution. This is likely partly due to the expected efficiency loss associated with the LDC allotment. Another factor is that households in different regions use energy in different ways and consume fuels with different levels of carbon-intensity. In some regions for example, electricity accounts for substantial proportion of cap-related costs; in other regions, natural gas or heating oil represents a greater cap-related cost. Further, in all regions, gasoline and indirect costs account for the highest proportion of cap-related costs, but the costs for these categories vary across regions. The lump-sum distribution approach would likely provide households with more flexibility to offset the cost increases specific to their situation, whereas an LDC allocation would favor those impacted the most by electricity price increases.

A primary trade-off of using the lump-sum household approach instead of the LDC allocation approach would be that non-residential electricity consumers (e.g., commercial and industrial entities) would not receive emission allowance value (unless they would receive it from other sources, such as an allotment for energy-intensive, trade-exposed industries). Although many of these entities have the option of passing through cap-related costs to customers, depending on their specific economic situation, entities' abilities to pass through costs would vary, but these issues are beyond the scope of this chapter.

APPENDIX. ADDITIONAL INFORMATION

Table A- 1. Comparison of Allowance Value Distribution Strategies from Select Cap-and-Trade Proposals Percentages Refer to Allowance Value (No-Cost Allowances or Auction Proceeds)

	108th Congress S. 139 (Lieberman)	**109th Congress S. 342 (S. 1151, S.Amdt.826)[a] (McCain/ Lieberman)**	**110th Congress S. 280 (Lieberman)**	**110th Congress S. 2191 (S. 3036)[b] (Lieberman/ Warner)**	**110th Congress H.R. 6186 (Markey)**	**111th Congress H.R. 2454 (Waxman/ Markey)**	**111th Congress S. 1733 (Kerry-Boxer)**
Date of Introduction	January 9, 2003	February 10, 2005	January 1, 2007	October 18, 2007	June 4, 2008	May 15, 2009	September 30, 2009
Legislative Activity	Referred to committee	S. 342 and S. 1151 referred to committee; S.Amdt. 826 offered (but not agreed to, 38-60) as amendment to H.R.6 on June 22, 2005	Referred to committee	Reported by Senate Committee on Env-ironment and Public Works May 20, 2008; considered (as S. 3036) by the Sen-ate during week of June 2, 2008	Referred to committee	Passed House June 26, 2009	Ordered reported by Senate Committee on Environment and Public Works November 5, 2009
Emission Allowance Distribution to Covered Entities	Percentages not specified EPA deter-mines perce-ntage of allowances to be provided	Percentages not specified EPA determines percentage of allowances to be provided at no cost to covered	Percentages not specified EPA determines percentage of allowances to be provided at	In early years of program: 19% to fossil fuel electric generators; 10% to energy-intensive sectors; 5% for early reductions;	In early years of program: 6% to manuf-acturers of"trade-exposed primary goods"	In early years of program: 13.4% to energy intensive, trade-exposed sectors; 3.5% to	In early years of program: 12.1% to energy intensive, trade-exposed sectors; 3.0% to merchant coal-

Table A-1. (Continued)

	108th Congress S. 139 (Lieberman)	109th Congress S. 342 (S. 1151, S.Amdt.826)[a] (McCain/Lieberman)	110th Congress S. 280 (Lieberman)	110th Congress S. 2191 (S. 3036)[b] (Lieberman/ Warner)	110th Congress H.R. 6186 (Markey)	111th Congress H.R. 2454 (Waxman/ Markey)	111th Congress S. 1733 (Kerry-Boxer)
	at no cost to covered entities	entities; rural electric cooperatives would receive allowances based on historical emissions	no cost to covered entities; rural electric coop-eratives would receive allowances based on historical emissions	2% to petroleum producers/importers; 1% to rural electric coops; 1% to HFC producers/ importers		merchant coal-fired generators; 2% to petroleum refineries (0.25% to small refineries); 1.5% for long-term power contracts	fired generators; 0.6% to petrol-eum refineries (0.4% to mid-sized and 0.9% to small refineries); 1.3% for long-term power contracts
Emission Allowance Distribution for Energy Consumers	Percentages not specified EPA determines percentage of allowances to be distributed for various policy obje-ctives; dete-rmination based on	Percentages not specified EPA determines percentage of allowances to be distributed for various policy objectives; determineation based on several economic and equity criteria, including efficiency and	Percentages not specified EPA determines per-centage of all-owances to be distributed for various policy objectives; determination based on sev-eral economic and equity	For program's entirety: 11% allocated to electricity and natural gas local distribution com-panies for energy consumer assist-ance	In early years of program: 58.5% to middle-and low-income households as tax credits and/or rebates	In early years of program: 30% to electricity LDCs for consumer benefit; 9% to natural gas LDCs for consu-mer benefit; 1.5% to states for heating oil consumers;	In early years of program: 26.7% to electricity LDCs for consumer benefit; 7.7% to natural gas LDCs for consumer bene-fit; 1.3% to states for heating oil consumers;

Table A-1. (Continued)

	108th Congress S. 139 (Lieberman)	109th Congress S. 342 (S. 1151, S.Amdt.826)[a] (McCain/Lieberman)	110th Congress S. 280 (Lieberman)	110th Congress S. 2191 (S. 3036)[b] (Lieberman/ Warner)	110th Congress H.R. 6186 (Markey)	111th Congress H.R. 2454 (Waxman/ Markey)	111th Congress S. 1733 (Kerry-Boxer)
	several economic and equity criteria, including efficiency and impact on consumers	impact on consumers; in S. 1151, 50% of this percentage must go to technology development	criteria, including efficiency and impact on con-sumers; 50% of this percentage must go to technology development			15% to low-income consumers In later years (after 2036): 34.7% to all households; 15% to low-income households	12.9% to low-income consumers In later years (after 2036): 37.7% to all households; 13.7% to low-income households

Source: Prepared by CRS.

Notes: The above table is not a comprehensive list of all cap-and-trade proposals, but includes select proposals from recent Congresses.

a. S. 342, S. 1151, and S.Amdt. 826 are identical in terms of emission allowance distribution.

b. S. 3036, introduced by Senator Boxer on May 20, 2008, is identical to the reported version of S. 2191, except that S. 3036 contains a budget amendment aimed at making the bill revenue-neutral. This would entail devoting a percentage of auction revenues—increasing from 6.1% in 2012 to 1 5.99% in 2031 and thereafter—to offset budget deficits that are projected to occur due to the cap-and-trade program. This bill was considered by the Senate the week of June 2, 2008.

End Notes

[1] Under the United Nations Framework Convention on Climate Change (UNFCCC), greenhouse gases are carbon dioxide (CO_2), methane (CH_4), nitrous oxide (N_2O), hydrofluorocarbons (HFCs), perfluorocarbons (PFCs), and sulfur hexafluoride (SF_6). Some greenhouse gases are controlled under the Montreal Protocol on Substances that Deplete the Ozone Layer, and are not covered under UNFCCC.

[2] The ability to limit GHG emissions already exists under various Clean Air Act authorities that Congress has enacted, a point underlined by the Supreme Court in an April 2007 decision, *Massachusetts v. EPA*. Although the current EPA Administrator has stated a preference for controlling GHG emissions through new legislation, the agency has begun to take actions that could lead to emission performance standards from particular sources. For more information on these developments, see CRS Report R40585, *Climate Change: Potential Regulation of Stationary Greenhouse Gas Sources Under the Clean Air Act*, by Larry Parker and James E. McCarthy; and CRS Report R40506, *Cars and Climate: What Can EPA Do to Control Greenhouse Gases from Mobile Sources?*, by James E. McCarthy.

[3] For further reading, see CRS Report RL345 13, *Climate Change: Current Issues and Policy Tools*, by Jane A. Leggettgett; CRS Report R40556, *Market-Based Greenhouse Gas Control: Selected Proposals in the 111th Congress*, by Jonathan L. Ramseur, Larry Parker, and Brent D. Yacobucci; CRS Report R40643, *Greenhouse Gas Legislation: Summary and Analysis of H.R. 2454 as Passed by the House of Representatives*, coordinated by Mark Holt and Gene Whitney; CRS Report R40896, *Climate Change: Comparison of the Cap-and-Trade Provisions in H.R. 2454 and S. 1733*, by Brent D. Yacobucci, Jonathan L. Ramseur, and Larry Parker.

[4] This could be accomplished by using the allowance value to offset reductions in other taxes, such as payroll or income taxes. For a further discussion of this and other allocation strategies, see CRS Report RL34502, *Emission Allowance Allocation in a Cap-and-Trade Program: Options and Considerations*, by Jonathan L. Ramseur.

[5] For more information see EIA, Electricity Basics 101, at http://www.eia.doe.gov/basics/electricity

[6] Calculated by CRS with data from the Energy Information Administration's 906/920 database, available at http://www.eia.doe.gov/cneaf/electricity

[7] The next largest contributor is transportation, accounting for approximately 28% of total U.S. emissions. EPA, *Inventory of U.S. Greenhouse Gas Emissions and Sinks: 1990-2007* (April 2009), Table ES-7.

[8] Anthony Paul *el al.*, *Compensation for the Electricity Consumers under a U.S. CO_2 Emissions Cap*, July 2008, Resources for the Future Discussion Paper. See e.g., EPA, *EPA Analysis of the American Clean Energy and Security Act of 2009, H.R. 2454 in the 111th Congress*, June 23, 2009, p. 11; and EIA, *Energy Market and Economic Impacts of H.R. 2454, the American Clean Energy and Security Act of 2009*, August 2009, p. 16.

[9] See CRS Report R40556, *Market-Based Greenhouse Gas Control: Selected Proposals in the 111th Congress*, by Jonathan L. Ramseur, Larry Parker, and Brent D. Yacobucci; CRS Report RL33 846, *Greenhouse Gas Reduction: Capand-Trade Bills in the 110th Congress*, by Larry Parker, Brent D. Yacobucci, and Jonathan L. Ramseur.

[10] For more information on these options see CRS Report RL34502 , *Emission Allowance Allocation in a Cap-andTrade Program: Options and Considerations*, by Jonathan L. Ramseur.

[11] See CRS Report R40896, *Climate Change: Comparison of the Cap-and-Trade Provisions in H.R. 2454 and S. 1733*, by Brent D. Yacobucci, Jonathan L. Ramseur, and Larry Parker.

[12] Both H.R. 2454 and S. 1733 would distribute free allowances to energy-intensive, trade-exposed industries. These industries may receive some of the allowance value distributed to electricity LDCs, but the bills direct the Environmental Protection Agency (EPA) (Sec. 764

in both bills) would take that into consideration when distributing allowances based on the facility's energy-intensive, trade-exposed status.

[13] Depending on their specific economic situation, entities' abilities to pass through costs would vary. This issue is beyond the scope of this chapter. See CRS Report R40914, *Climate Change: EU and Proposed U.S. Approaches to Carbon Leakage and WTO Implications*, by Larry Parker and Jeanne J. Grimmett.

[14] Testimony by Douglas W. Elmendorf (Director of the Congressional Budget Office) before the Senate Committee on Finance (May 7, 2009).

[15] Ibid.

[16] Anthony Paul *et al.*, *Compensation for Electricity Consumers under a U.S. CO2 Emissions Cap* (July 2008), Resources for the Future Discussion Paper.

[17] Rich Sweeny and Dallas Burtraw, *The Effects on Households of Allocation to Electricity Local Distribution Companies* (May 19, 2009).

[18] Interagency Report (in response to a request from Senators Bayh, Specter, Stabenow, McCaskill, and Brown), *The Effects of H.R. 2454 on International Competitiveness and Emission Leakage in Energy-Intensive Trade Exposed Industries* (December 2009), p. 35 (footnote 41), at http://www.epa.gov/climatechange/economics/pdfs/ InteragencyReport_ Competitiveness&EmissionLeakage.pdf.

[19] Douglas W. Elmendorf (Director of the Congressional Budget Office), Testimony before the Senate Committee on Finance (May 7, 2009).

[20] U.S. Bureau of Labor Statistics, Consumer Expenditure Surveys, Table 45, at http://www.bls.gov/cex/.

[21] This factor relates to the price elasticity of demand for electricity and gasoline, a subject that is beyond the scope of this chapter.

[22] Rich Sweeny and Dallas Burtraw, *The Effects on Households of Allocation to Electricity Local Distribution Companies* (May 19, 2009), Resources for the Future.

[23] Dallas Burtraw (Resources for the Future), Testimony before the Senate Committee on Finance (August 4, 2009).

[24] Ibid.

[25] Many consider CO_2 emissions from biomass sources as practically neutral, because biomass sources take in carbon dioxide during their growing cycle and release it when burned, but these issues are subject to intense debate. See CRS Report RL34059, *The Carbon Cycle: Implications for Climate Change and Congress*, by Peter Folger.

[26] Dallas Burtraw, et al., *The Incidence of U.S. Climate Policy: Alternative Uses of the Revenue from a Cap-and-Trade Program* (June 2009), Resources for the Future Discussion Paper 09-17-Rev; and Kevin Hassett, *et al.*, "The Incidence of a U.S. Carbon Tax: A Lifetime and Regional Analysis" (2009) *The Energy Journal*, Vol. 30, No. 2.

[27] For a discussion of these studies and their differences, see CBO, *Two Recent Studies of Regional Differences in the Effects of Policies That Would Price Carbon Dioxide Emissions* (July 9, 2009).

[28] The NBER study did not provide estimated dollar value impacts, but they would likely be comparable to RFF data as the assumed carbon prices were relatively similar.

[29] See CRS Report R40809, *Climate Change: Costs and Benefits of the Cap-and-Trade Provisions of H.R. 2454*, by Larry Parker and Brent D. Yacobucci.

[30] Power plants often consist of multiple generating units. For example, a single plant site may have one unit that burns coal and a second unit that burns gas. These units may have different ownership. Prior to 2008, EIA collected data at the plant level and did not distinguish which generating unit at a plant site was consuming a particular fuel. With advent of the new EIA-923 survey in 2008, this data is now collected at the generating unit level. A now discontinued survey, the EIA-767, collected unit level information for a limited set of power plants.

[31] Prior to January 2008, the data was collected on two predecessor forms, the EIA-906 and EIA-920 surveys.

[32] A power sales contract may identify the nominal source of the power. However, this does not mean that the electricity actually comes from the specified power plant. On a given day, a generator might choose for economic or operational reasons to run a different unit. As long as the contract price is charged and there are no transmission congestion implications, the power buyer is generally indifferent to the actual source of the electricity.

[33] Both H.R. 2454 and S. 1733 have such provisions.

[34] EPA, Emissions & Generation Resource Integrated Database, at http://www.epa.gov/cleanenergy.

[35] See e.g., Testimony of Jeffry E. Sterba (on behalf of the Edison Electric Institute), House Committee on Energy and Commerce Subcommittee on Energy and Environment, Hearing on Allocation of Emissions Allowances (April 23, 2009.)

[36] The states include Illinois, Indiana, Kentucky, Michigan, Missouri, Ohio, West Virginia, and Wisconsin. See Dallas Burtraw, et al., *The Incidence of U.S. Climate Policy: Alternative Uses of the Revenue from a Cap-and-Trade Program* (June 2009), Resources for the Future Discussion Paper 09-17-Rev.

[37] For a discussion of allowance price estimates of H.R. 2454 (relevant also to S. 1733), see CRS Report R40809, *Climate Change: Costs and Benefits of the Cap-and-Trade Provisions of H.R. 2454*, by Larry Parker and Brent D. Yacobucci.

[38] See e.g., Chad Stone, *Holding Down Increases in Utility Bills Is a Flawed Way To Protect Consumers While Fighting Global Warming* (June 3, 2009 revised), Center on Budget and Policy Priorities; Dallas Burtraw, *Testimony before the Senate Committee on Finance* (August 4, 2009), Resources for the Future.

[39] S. 1733 (Section 772(b)(5)(C)) and H.R. 2454 (Section 783(b)(5)(C)).

[40] Sebastian Rausch *et al.*, *Distributional Impacts of a U.S. Greenhouse Gas Policy: A General Equilibrium Analysis of Carbon Pricing* (November 2009), MIT Joint Program on the Science and Policy of Global Change.

[41] Frederick F. Butler, President, National Association of Regulatory Utility Commissioners, Letter to Chairman Boxer and Ranking Member Inhofe (October 28, 2009).

[42] Ibid.

INDEX

D

E

P

R

S